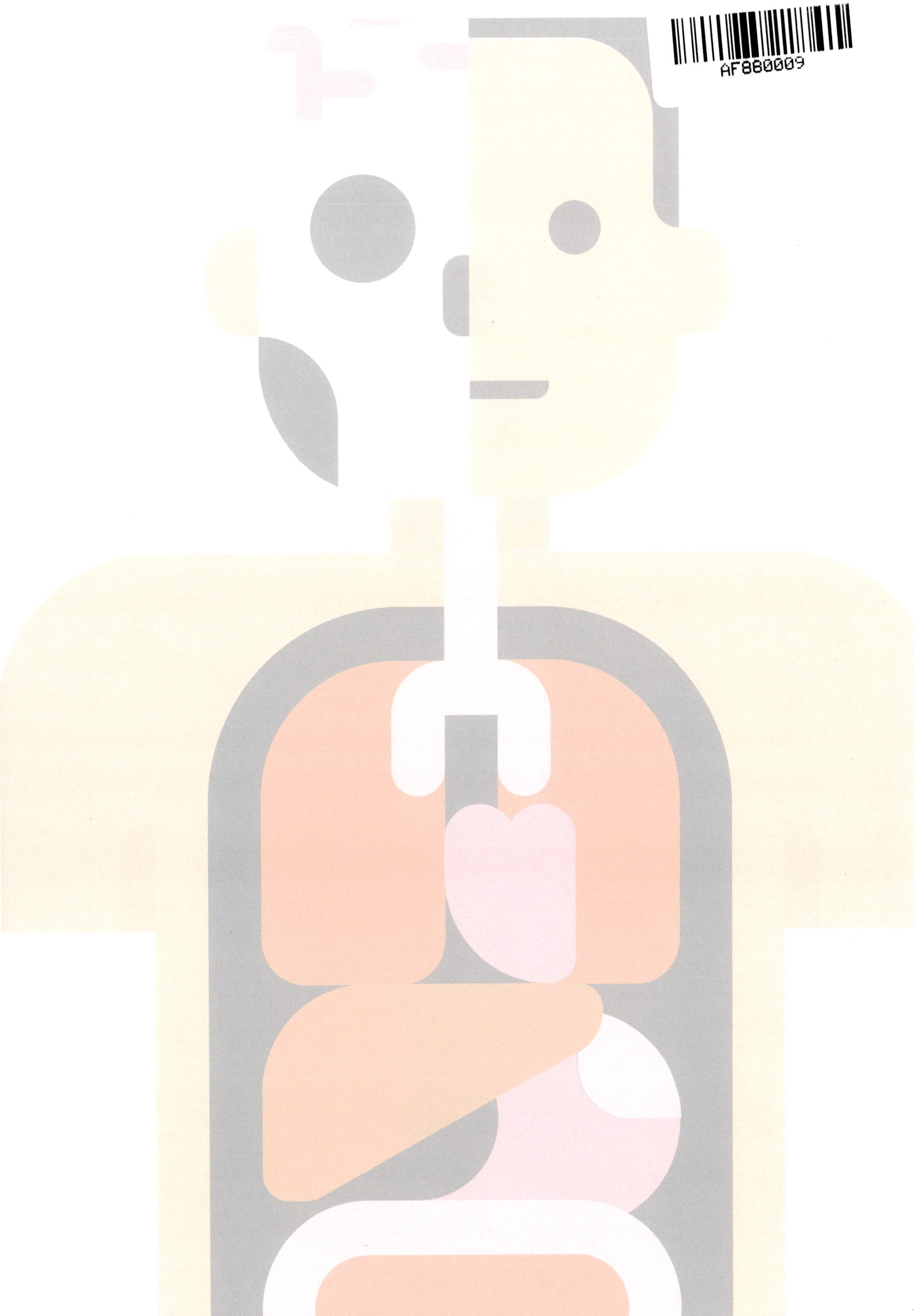

TAPPING FOR TRAUMA

TAPPING FOR TRAUMA

A CLINICIAN'S GUIDE TO HOW DUAL AWARENESS THERAPIES WORK

Published by *Barna Books* an imprint of Barna Counselling & Consulting.

Melbourne, Australia.

 All cases outlined in this book have been altered to protect client privacy. Where dialogues are referenced, all care has been taken to reproduce the essence of the conversation, if not verbatim statements. This book is not intended to be a substitute for advice from a trained professional.

Cover design by Barna Books via Visme

Book format by Barna Books via Visme

ISBN (paperback): 978-0-6456654-1-3

ISBN (eBook): 978-0-6456654-0-6

www.barnacc.com

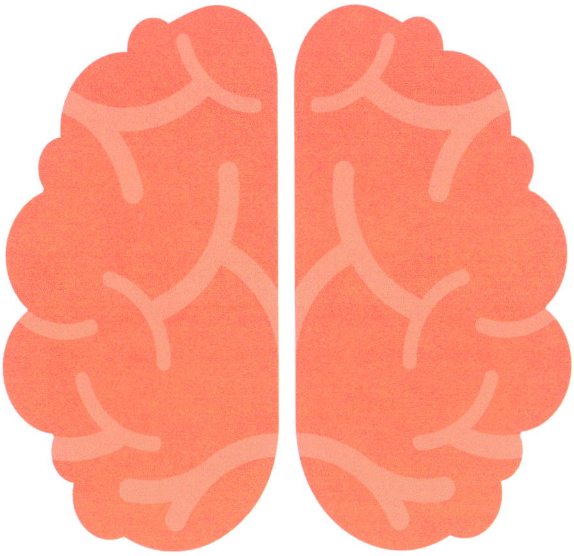 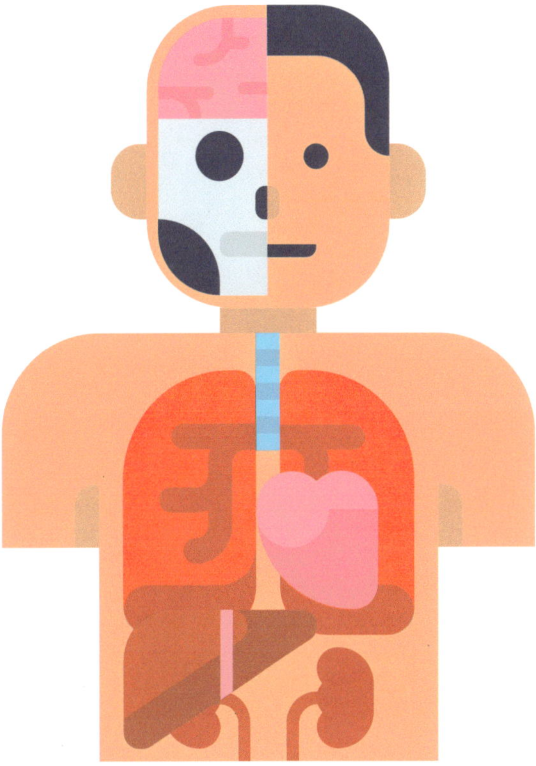

TABLE OF CONTENTS

AUTHOR'S NOTE

AUTHOR'S NOTE

Therapy Works!

> ***Did you know that psychotherapy works for ~80% of people who access it?*** [1]
> ***This means ANY type of therapy—cognitive, behavioural, somatic, existential, neurobiological, spiritual—has its utility.***
> ***So how do dual awareness therapies (like tapping) stack up?***
> ***The aim of this book is to find out.***

This book is for anybody who wants to know how dual awareness therapies such as the **Emotional Freedom Technique** (**EFT**), **Radical Exposure Tapping** (**RET**), and **Eye Movement Desensitization and Reprocessing** (**EMDR**) work at a neurobiological level. The dual awareness therapy I'm trained in is Laurie MacKinnon's RET—something I've published on alongside the psychologist, Dr Denis O'Hara in 2019 [2].

However, this book is NOT about any one therapy in particular. Rather, I seek to explain how factors observed in dual awareness therapies might work at a neurobiological level. If you're anything like me and want to understand the mechanics behind what you're doing, to feel confident about explaining to your clients what happens when dual awareness, novelty, and human connection are made therapeutic, this is for you.

Caveat: *I am NOT a neuroscientist—I am a psychotherapist. However, I have taken pains to ensure the information presented is accurate and useful. I hope you find the content as helpful as I have putting it all together.*

Emil Barna

BA (Psychology), Grad Dip (Counselling),
Master (Counselling and Psychotherapy)
Psychotherapist | Podcaster | Writer | Founder
Barna Counselling & Consulting
Barna Books
www.barnacc.com
emiliobarna@gmail.com

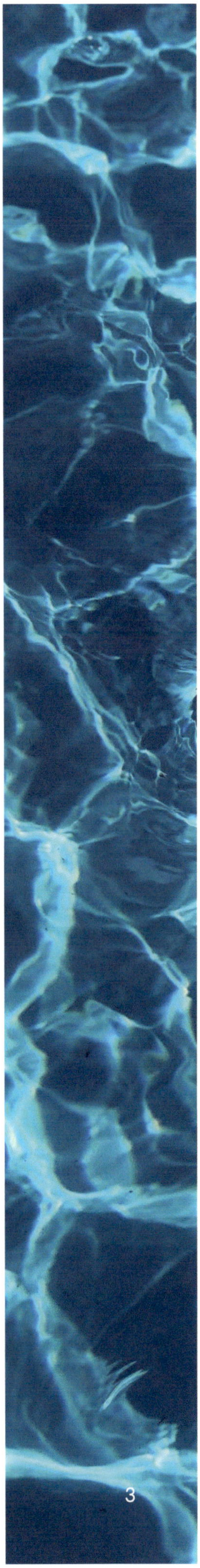

AN INTRODUCTION

> ***"All truth passes through three stages: First, it is ridiculed. Second, it is violently opposed. Third, it is accepted as self-evident."***
>
> —Arthur Schopenhauer

A Brief History of Tapping

A while back, I listened to a podcast where the late psychologist, Francine Shapiro (1948-2019), was being interviewed. She made a comment that's stuck with me ever since.

It went something like this:

"At the beginning, psychological treatment began with the **biological paradigm** with the German neurologists of the day. Then came the **psychodynamic tradition**, which aimed to uncover deep-seated traumas and conflicts that flooded everyday life. The **behaviourist tradition** came next, focusing on conditioning and pairing actions to evaluate responses. After **behaviourism**, the **cognitive revolution** took centre stage with the main focus on dissecting and challenging negative thoughts."
Shapiro went on: "We're the new kids on the block with **EMDR**. It's only natural that people would be skeptical."

I begin with EMDR because of its frank novelty in the realm of trauma treatment. It was vastly different to those therapies preceding it. Tapping and dual awareness therapies in general (i.e. those that guide a person to impart awareness between many things at once, keeping one foot in the past and another in the present) begin with the same premise: we've got to do something different than what we're used to. Enter the current wave in trauma treatment—one that utilises the body as much as the mind to help a person heal from pain.

It may come as a surprise that following a simple walk in the park, a revolution in trauma treatment began. It was 1987, [3][4] Shapiro was stressed. She was a psychology graduate, and she hadn't quite settled on a research topic. In need of fresh air, she decided to take a walk and clear her head. Heading off to a local park lined with large trees, she let her eyes wander. As they did, she noticed they would dart from left to right, then back again. Her focus? The oaks on each side of the path she was treading. At the same time, she began to feel her stress decrease.

This is weird, she thought and continued to scan the tree lines. *Why less stress?* She stopped, smiled, and nodded to herself. Perhaps there was something in this dual awareness?

From this humble experience, Shapiro chose to research eye movements in the reduction of stress. A therapy emerged and was called EMDR. Today, EMDR is an eight-phase treatment approach "that includes **affect regulation** components and desensitisation of a target memory via **bilateral stimulation** once somatic, visual, and cognitive cues are ascertained." [5] It has since become a frontline treatment for many psychiatric conditions, most notably **posttraumatic stress disorder** (**PTSD**). [6] [7] [8] [9]

Similar to EMDR, **Emotional Freedom Techniques** (**EFT**) arose unexpectedly at the poolside of the late psychologist Robert Callahan (1930-2013). It was 1979. [10] Callahan was treating a lady called Mary for a water phobia. At the time, his interest in **alternative therapies**, specifically **acupuncture**, had grown from a fledgling flirtation to a deep-seated admiration. Mary was struggling. Sitting beside her (and thinking on his feet), he decided to guide her in a relatively novel process of physically tapping a part of her body (her abdomen) associated, according to Chinese medicine, with the stomach **meridian**. Callahan's instinct was informed by Mary saying that she always felt her discomfort with water in her stomach. He was surprised to see that through this process, Mary's phobia of water evaporated—a finding not dissimilar to Shapiro's eye movements and how they paved the way for a felt sense of calm and collectedness.

Callahan's informal experiment led him to develop different 'tapping algorithms' for various triggers. It was dubbed: **Thought Field Therapy** (**TFT**). [11] [12] Gary Craig—a student of Callahan's and a former engineer—thought there must be a better way than to memorise a million-and-one tapping algorithms for any given issue. He decided to simplify the tapping procedure and devised one easy-to-remember algorithm. He called it EFT. [13] EFT's main goal was to reduce negative affect (observable emotions) associated with issues such as anticipatory anxiety, chronic pain, traumatic memories, and generalised anxiety. This was done by bringing to mind a target issue whilst physically tapping on various locations throughout the body. EFT asks the person to recite negative self-statements aloud as reminder phrases (e.g. "I feel bad", "I'm no good", etc.) while simultaneously tapping and reciting self-affirmations (e.g. "I'm here now. I'm okay.").

It was Craig's approach that gained traction and is now considered an effective, evidence-based approach to treating various mental health conditions. [14] [15] [16]

RET was developed by the family therapist Laurie MacKinnon to combine what she thought were the most effective elements of EMDR and EFT to create an economical, time-efficient, and versatile approach for the treatment of trauma and emotional reactivity. [17] [18] MacKinnon was after a 'silver bullet' to help her clients overcome fear-conditioning. The differences between RET and EFT are not stark, and one could safely see both are quite similar in practice. Though, an argument could be made that RET is more structured.

First, the RET protocol guides the client in safely recalling a target memory, related thoughts, somatic sensations, imagery cues (pictures that come up upon recall), and associated distress. In short, it asks you to describe the memory. The clinician then guides the client in using the EFT tapping procedure while reciting stressful cues until **SUD** (**subjective units of distress**) reduce in intensity. The memory is 'reprocessed' when SUD are close to 0/10, and a new memory (or image, or thought, or feeling) is 'installed', replacing the old one. Any computer engineer would notice the reprogramming parallels in this exercise. This installation phase appears transtherapeutic in that many modalities help clients reduce the heat associated with a traumatic memory, replacing it with a more adaptive response.

Theoretically, RET utilises the explanatory power of the **working memory hypothesis** (originally attributed to EMDR) to back up a potential mechanism of action for the RET protocol. [19] This theory posits that as one person maintains **dual awareness** between an emotional trigger and their present experience (e.g. thinking about a horrible memory while simultaneously focusing on the room one is in—the 'here-and-now), they overload their **working memory** (i.e. the cognitive space that retains short-term verbal, visual, and attentional information, allowing manipulation between them [20]). This leads to less of an emotional punch when the memory becomes reactivated. [21][22]

An astute clinician would observe this type of therapeutic intervention—shifting attention between a noxious trigger and another bit of information, thus attenuating the initial reaction—is found in many trauma therapies, from Cognitive Processing for Trauma to Somatic Experiencing. In terms of explanatory mechanisms, MacKinnon seeks to bypass the scientifically speculative claims at the heart of energy psychology approaches which contend that therapeutic change rises from the purposeful tapping of various meridian points scattered throughout the body, preferring an explanation consistent with that of EMDR. However, in doing so she ignores the fact that rarely (even in EFT circles) do individuals ascribe the process of change to meridian tapping and there are more nuanced explanations that are pursued (see Church in the **Endnotes** for more).

One might also argue that MacKinnon conflates one theory of action (attributed to EMDR) with RET to explain how her modality works. Just because they are similar modalities does not necessarily mean they work in the same way. They might, but confluence in therapy does not necessitate convergence. It is for this reason that I seek to provide a neurobiologically harmonious explanation of a *variety* of factors that may be at work in treatments that require dual awareness. But first, let's consider the kinds of explanations that have been proposed for these types of memory reprocessing techniques until now.

1. Imaginal exposure. This theory posits that change is achieved by slowly exposing a person to psychologically noxious (harmful) stimuli until desensitised. [23] Just as an initially scary movie might make you jump out of your skin, the more it's watched, the less it might make you shiver. Monsters lose their edge with each confrontation.

2. Bilateral hemispheric stimulation. This theory posits that stimulation of both hemispheres of the brain allows for interhemispheric (i.e. between-hemisphere) integration. The reduction of negative emotion occurs once both hemispheres are stimulated. [24] As one switches between different elements of a memory (e.g. from an image [the domain of the right hemisphere] and associated thoughts [the domain of the left hemisphere]), troubling emotions subside—just like reciting a maths equation after thinking of something stressful shifts the way you feel about the more emotional event. Interhemispheric integration poses that something of the sort happens as one focuses on varying elements in the here-and-now.

3. Rapid eye movements (**REM**). This theory proposes that quick eye movements from side to side allow for imaginal integration and affect reduction in a similar way that REM sleep allows the mind to process the day's events while asleep. [25] Just as dreams make their way into our consciousness when we're not purposely avoiding them, REM theory proposes that stimulating the left and right hemispheres through REM can soften the memory.

4. Adaptive information processing model (**AIP**). AIP posits that memories stored in an unprocessed, top-heavy manner are recalled and re-activated at a neural level—somatically, cognitively, imaginally. With treatment, these memories are reprocessed into a form that allows for positive re-learning. [26] [27] [28] The AIP model posits that forgotten or suppressed memories can be re-experienced as flashbacks during stressful situations. However, once broken down into their component parts (what a person sees, hears, thinks) and evaluated therapeutically, there's the opportunity for healing.

In any case, there are perhaps as many theories as there are theorists, but I hope I've given you some understanding of what's currently being thrown around. From now, we'll shift gears and ease into the neuroscience of it all.

The human brain comprises of a complex labyrinth of interconnected **neurons** for its daily operation—around 86 billion to be precise. [29] [30] These neurons ignite in a flurry of **electrochemical activity**, sending and receiving messages every millisecond of every day. [31] In addition, there are 80+ billion **glial cells** supporting neurons by strengthening neural connectivity, insulating neural membranes, and even creating new brain cells. [32] In essence, the brain is made up of ~160 billion cells that work in unison to help us do what we need to at any given moment.

It's a gargantuan task for any researcher to answer to how any single therapeutic modality or protocol is involved in altering specific neural circuitry. In fact, it may be an impossible task. Even the most advanced brain-scanning systems we have—**SPECT imaging**, **fMRI**, and even the **CAT scan**—cannot provide more than a glimpse into the brain within a given time slot. Perhaps the best one can do is look towards neural correlates and, where possible, identify the mechanisms at play when a therapeutic intervention works. [33] This is not always easy.

In this book, I aim to highlight what *might* be happening at a **neurobiological** level within the context of tapping and other dual awareness protocols, using neuroscience literature as a basis for understanding. The psychiatrist's Bible—**Diagnostic and Statistical Manual of Mental Disorders** (**DSM-5**; now in its fifth edition, recently revised [34])—will not form the basis of my commentary. Instead, I'll refer to more broadly defined stressor-related and traumatic experiences, exploring extant literature in understanding how these experiences are encoded within the human brain and, more importantly, how one can overcome them.

However, my utilisation of these neurobiological underpinnings is speculative. To date, there is no published study I am aware of that confirms what precisely occurs in the brain while a person is undergoing dual awareness therapy. To bridge this gap, I shall rely on what is more or less accepted in the neuroscientific field before applying it specifically to the clinical sphere.

So with that preamble, let us continue our journey into the human brain.

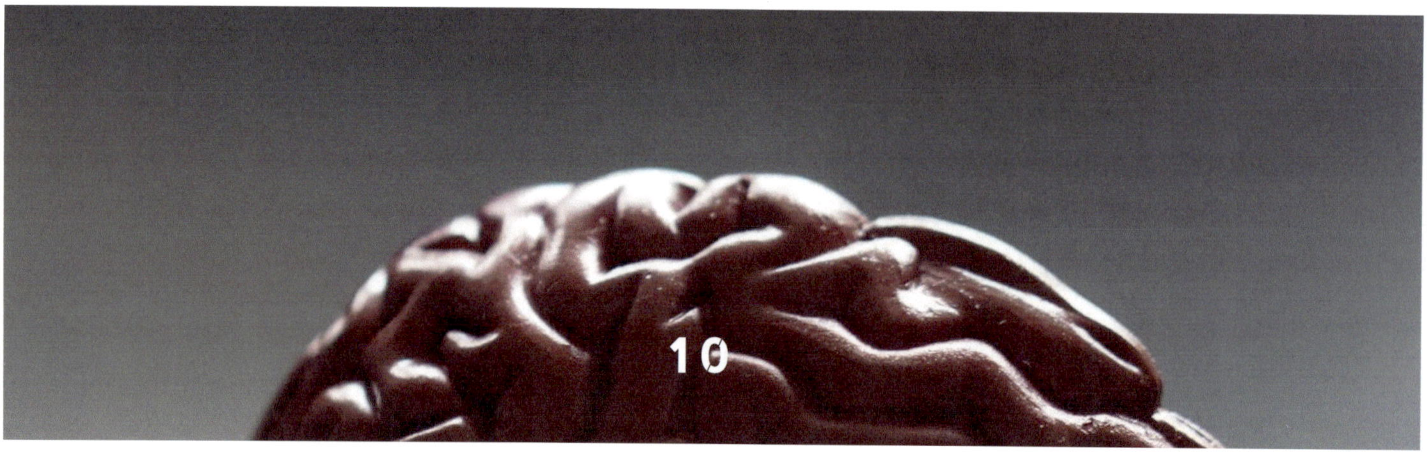

1

HUMAN EXPERIENCE & THE BRAIN

Encoding Human Experience

What one calls **memory** can also be understood as the repetitive firing of closely linked neurons in the brain. Of course, that's not *all* memory is just like love is not only the excretions of various neurochemicals and pituitary hormones when we get *that feeling*. But it helps, at least for our purposes, to discuss these goings-on so that we can form a foundation for understanding how various therapies work.

Neural firing forms new **neural pathways** (also called **neural networks**) within the human brain. The brain is an associational machine, linking patterns of behaviour and experiences with preferential outcomes so that we can repeat them. In the case of memories, experiences show up in our brain using the same neural networks that were involved when initially encoded. [35] This neural firing pattern was first proposed by the late neuropsychologist Donald Hebb (1904-1985). It was later termed the **Hebbian principle**, positing that as one neuron is activated in close proximity to another, the two become closely linked and continue to fire together in the future. [36] [37] This principle was popularised with the work of the neurobiologist Carla Shatz in her clever mnemonic, *what fires together, wires together* [38]—a rhyme scientists and clinicians still use today. [39]

With memory, experiences that are especially palpable become emotionally charged. The brain recognises these experiences as important and lays down the neural circuitry required to be activated in the future. This potentially disastrous series of events occurs vociferously in the case of individuals who have experienced incredibly stressful or traumatic events—events that often haunt them in their waking hours, influencing their view of themselves, others, and the world at large.

Thankfully, treatments are available that can potentially reverse this encoding process and lead to remarkable results.

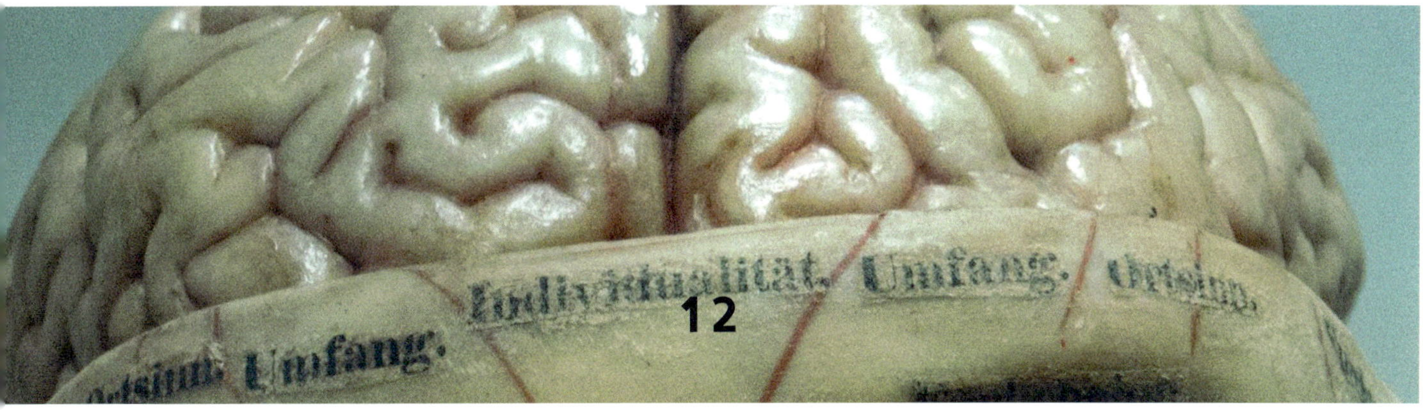

Stress, Trauma & the Human Brain

Many conditions that mental health professionals treat are stress-related.

In Australia, of over 19.6 million Australians surveyed in the most recent *National Study of Mental Health and Wellbeing* [40], 43.7% (~8.6 million people) had experienced a mental disorder at some time in their life. About one in five people (21.4% or 4.2 million people) had a diagnosable mental disorder over the past 12 months (a small increase from the previous survey in 2007).

PTSD went from the most diagnosable mental health condition in 2007 to the second most diagnosable in 2021. The first? Social Phobia—an understandable finding when one considers COVID-19 and the resulting lockdowns. We are an incredibly stressed society.

These findings deserve attention, but it is attention I cannot give here except to say that clinicians must continue their focus to address stress-related conditions for the Australian demographic, perhaps even more than they already do.

These statistics don't come as much of a surprise to the astute observer. Putting the diagnostic criteria of the DSM-5 aside and refocusing on the encoding of human experience, one can understand it is not so shocking to see that stress-related issues like PTSD are the most diagnosed—they are encoded most prevalently.

Let's consider this for a moment ...

Stress

"You know that feeling when you catch your foot in a crack in the concrete, and you're about to trip over, but then you don't—that's what it's like for me, all the time."

"It's like you're staring at a train—you're on the tracks—then you jump aside at the last minute, huffing and puffing. That's what the feeling's like for me."

The speech bubbles on the previous page reflect what my clients have said about their experiences of stress. Although there are variations, they share commonalities. Across the board, when a person is threatened, the initial response is to **arrest** (stop and scan for danger) **and orient** towards the potential threat. [42]

Suppose there's reason to fear, and the threat was not just a false start; in that case, the sympathetic nervous system (one of the two branches of the autonomic nervous system—a body system involved in maintaining automatic functions like heartbeat and digestion) kicks into high gear. It initiates two of the most common responses to threat: **fight-or-flight**. [43] This response reflects an adrenaline pump—your body tenses up and you brace for danger.

Each of us has our unique response. Generally—depending on prior conditioning and physical ability—you either orient toward the threat with the carnal goal of fighting it off, or you'll take flight.

The human stress response is not as clear-cut as if-this-then-that. No. Each person has their own nuanced relationship with stress. But there are similarities. Let's consider rodents for a moment as they share the basic anatomical brain organisation as mammals in terms of threat detection and response. If a rat is threatened, its instinct is to take off. Only when it's cornered will it fight back. Many humans are the same—only when they feel cornered (literally or figuratively) will they bite the head off their opponent.

In addition to the fight/flight responses, mounting research has also identified a **freeze response**. This freeze response is a common precursor to flight, resembling the act of halting, camouflaging into the background or blending in. [44] [45] [46] [47] Freeze can also occur following fight-or-flight if you've exhausted all other options and still cannot escape. [48]

A few years back, someone rammed into the back of my SUV while I was on my way to work. Anybody who has been in a car accident knows how involuntarily tense your body gets when something like that happens—it's as if you're frozen. The funny thing is, as I first wrote this very paragraph—and for the longest time after the accident when I travelled on the same stretch of road—my lower back received a spike of pain. This itself, as many somatic therapists would argue, is the result of the freeze response locking itself in the body. When **sympathetic arousal** cannot protect a person from the threat, the body may initiate the **fall response** through dissociation or the actual loss of consciousness. This happens by preserving energy and protecting the organism from mortal injury. [49]

The fall response is the body's last-ditch effort to protect itself by preserving and, at times, slowing down basic functions (e.g. heartbeat, breathing rate, blood clotting abilities, digestion) to survive. Clinically, this can look like the person is zoning out, feels outside their own body, cannot pinpoint how they feel, cannot register physical pain as readily as when not in this state, might fold into themselves, and experience severe lethargy and (less frequently) fainting.

It might not feel like it, but the fight-flight-freeze responses, as intense and uncomfortable as they feel at the time, are better states to be in from a stress viewpoint, given the body still feels like it can tackle the threat. Not so with fall.

So what?

Now that we've got a beginner's understanding of the initial stress response, let's ask what happens to a person when they're constantly under stress or traumatised. In a previous book, [50] I tackled this topic at length. For a brief overview of what happens during chronic stress, turn to the next page.

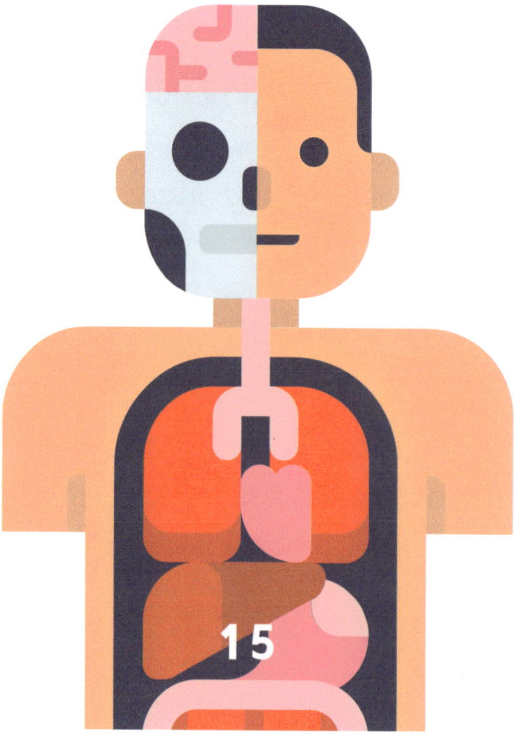

HUMAN

STRESS RESPONSE

Heightened blood pressure (***hypertension***) accelerates the body's push to send blood where it needs to go, thereby making it crash against the blood vessels at higher volocities

Higher blood glucose and fat levels provide the body with more useable energy during times of increased stress.

Higher metabolic rate means a heightened energy release leading the body to sweat in order to cool it down for sustained energy

Blood gets thick and clots much faster than usual. This is especially important if you suffer an injury

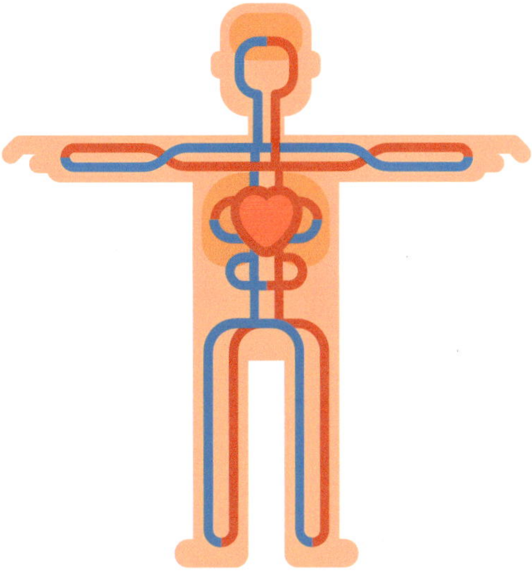

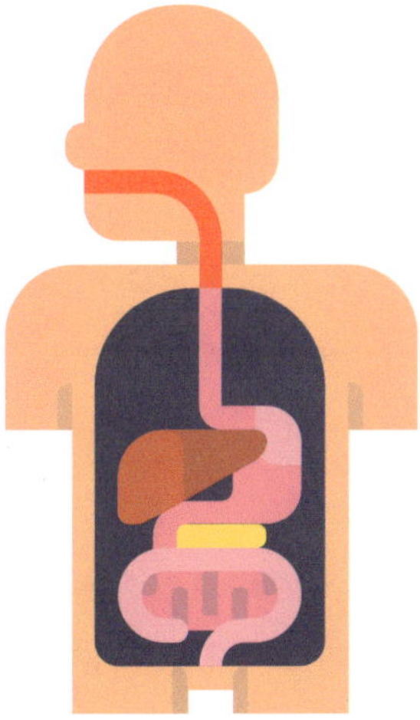

More intense heartbeat (***heightened cardiac contractility***) leads to more blood being sent to where it's needed most

The gastrointestinal system slows or shuts down so that the body can use whatever blood is necessary to tackle the threat

The immune system pumps out inflammatory chemicals such as histamine and serotonin which help parts of the body that are injured so that any associated damage doesn't spread too far

Eyesight becomes more acute and sharp so that you can pick up easily missed details in the environment

As you can see, the human stress response has remarkable utility. Why would it not? These adaptations have accumulated over time throughout our evolutionary history to maximise reproduction and survivability. And remember, the functions I listed earlier are only a few of what the body is known to do. I haven't mentioned the complex hormonal release that includes the flooding of dissociative painkillers when under duress, changes in musculature that lead to protective responses, and flinching capabilities that occur below conscious awareness.

Trauma

Trauma is *any* event perceived as overwhelming at the time of its happening and that causes lasting physiological, neurobiological, psychological, interpersonal, and spiritual problems long after it has ceased.

In my clinical practice, trauma presents itself in many ways, but overwhelmingly it presents in the body. From a thumping heart to a dropping gut, a dripping forehead to clenched fists, trauma affects the body in idiosyncratic and nuanced ways. The main thing separating the traumatic stress response from 'ordinary' stress is the person's (or organism's) sense of *inescapability*—they feel cornered, cannot escape, take control, or overcome the threat. [51] The stress response is then 'locked' in the body and it is very difficult to return to equilibrium. [52]

I mentioned earlier that the definition of trauma need not be restricted to the categorical imperatives of the DSM-5-TR in the form of PTSD, however useful this nomenclature is for the diagnosis and treatment of traumatic incidents. [53] [54] [55] [56] [57] One can accept the utility of this transdisciplinary understanding of traumatic re-experiencing without conceding that this is the only way to understand stressor-related conditions. As Peter A. Levine, the originator of the trauma therapy Somatic Experiencing, put it: *Trauma is trauma, no matter what caused it.* [58] And also, from the lips of the well-known trauma therapist and researcher Bessel van der Kolk: *The hallmark of trauma is inescapability.*

The Human Brain

The human brain operates through an infinite complexity of imperatives that respond to the environment in the case of a threat. Within a thousandth of a second (quicker than a housefly's wing flap), a series of neurons fire in response to a potential threat. This forced

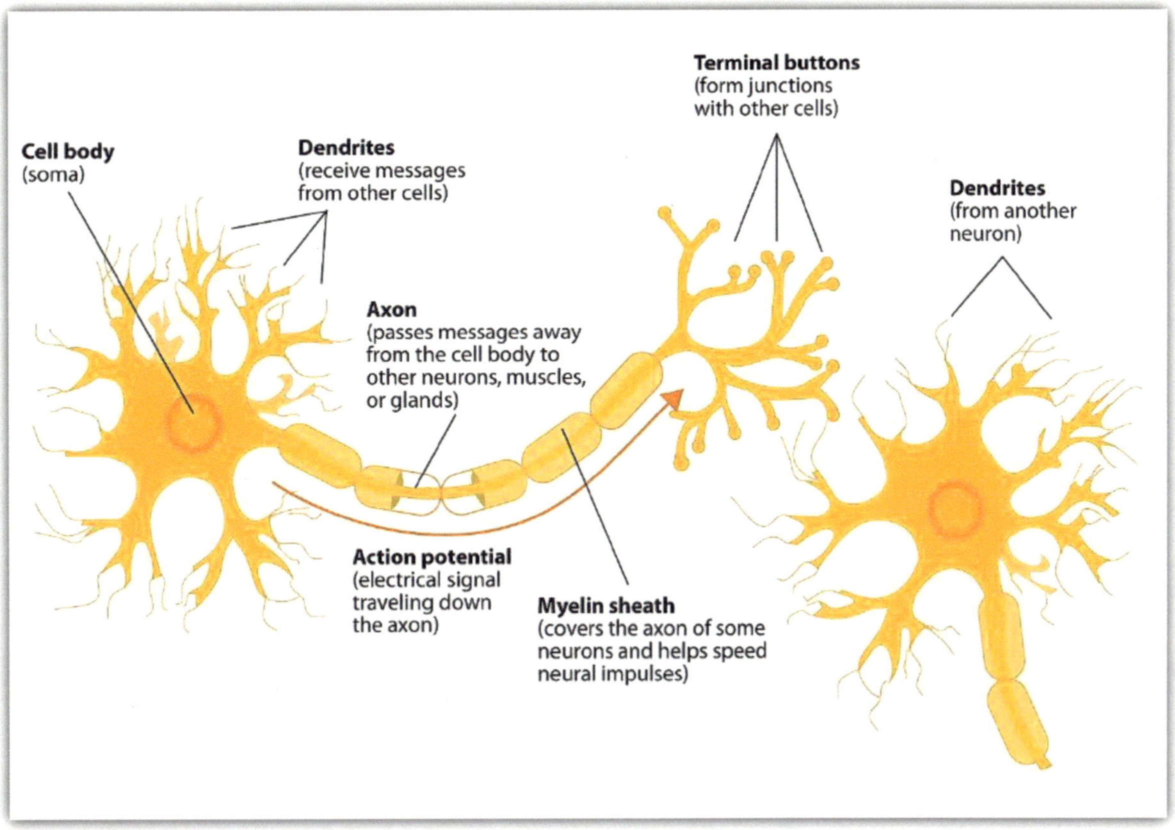

Fig. 1. The anatomy of a neuron firing.
Source: Jennifer Walinga

action initiates a domino effect that uses the time-honoured wisdom of the **thalamus** (a bunch of brain cells clumped atop the **brainstem** to form the cranially-centred neural region responsible for sifting and sending sensory data—sight, sound, feel, smell, taste—to where it's needed most). [59] [60]

The thalamus alerts the **amygdalae** (commonly referred to as the **amygdala**), two neural clusters located within the **limbic system** (a region of the brain responsible for emotional activation, fear, memory, and attention). [61] The amygdala awakens the **hypothalamus** (an almond-sized structure located just beneath the thalamus, and just above the roof of your mouth, responsible for automatic somatic processing, including the preparation and response to danger). [62] Like a chain reaction, the hypothalamus signals the **pituitary gland**. This pea-sized extension hangs beneath the hypothalamus and releases select stress hormones to tackle perceived threats. This process then kicks off the fight-flight-freeze instinct by facilitating relevant **endocrine glands** throughout the body to release the required hormones. [63] The mind-bender is that all this happens between three to f-

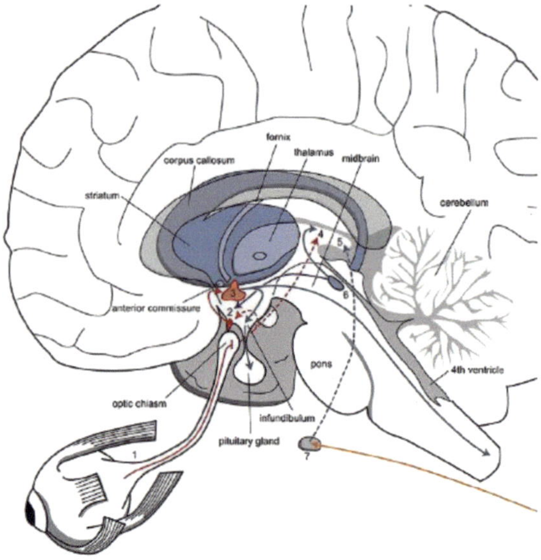

Fig. 2. Various brain regions, including the thalamus, pituitary gland, and the brain stem (latter not labelled, but appearing below the midbrain, toward the spinal cord.
Source: Anton J. M. Loonen and Svetlana A. Ivanova

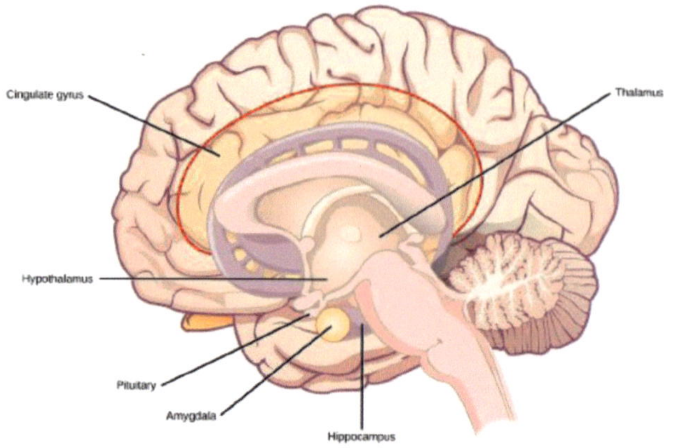

Fig. 3. Above are various brain regions found within the limbic system, including the hypothalamus, thalamus, pituitary gland, hippocampus, and amygdala. Keep in mind that these structures have their twin in the opposite hemisphere.
Source: CNX OpenStax

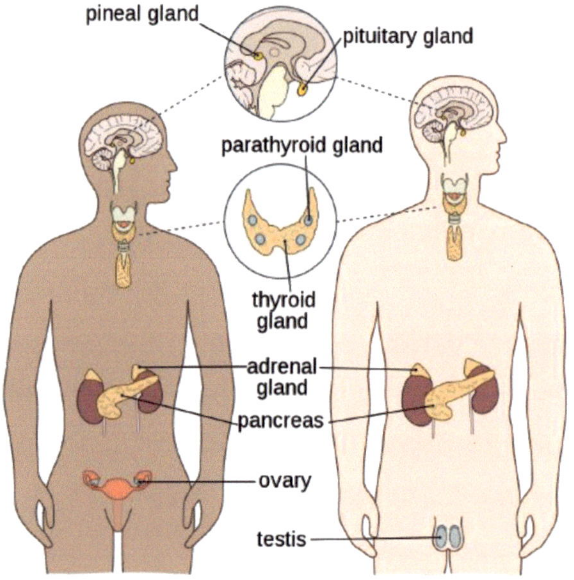

Fig. 4. The endocrine system, with the adrenal glands labelled.
Source: OpenStax & Tomáš Kebert & umimeto.org

our hundred times quicker than a person's conscious awareness of any such threat even being there. [64] [65]

Your body reacts even before you're aware there's a threat!

In the case of traumatic experiences (and even stress in general), the **adrenal glands** (located atop the kidneys) are triggered by the pituitary gland to secrete an influx of **cortisol** an-d **adrenaline** to prepare the body for action. This process, in turn, blocks the function of the **hippocampi** (commonly referred to in the singular—**hippocampus**—clumps of neurons located within the **limbic system**; refer to **Fig. 3**). These neural structures, cortically close to the amygdala, are responsible for the formation, storage, and retrieval of new memories. [66] They are also fundamental for integrating experiences into a cohesive memorial whole. [67] Clinically, this can mean a person finds it difficult remembering what must be done, forgets essential details of what's happened to them and, for some, forgets who they are when under significant stress—an experience termed **depersonalisation** (literally, forgetting oneself).

Consider the example from the clinical psychologist Jordan Peterson. He said, "If you're in enough pain, you're no longer self-conscious" [68] when describing his physical and mental anguish relating to **akathisia** (a debilitating movement disorder caused by psychopharmacological intervention) to *The Sunday Times*.

When escape is not an option, these neural and hormonal responses perpetuate the problem, pushing the person into hypervigilance and threat anticipation long after the stressor has passed. [69] The brain, in turn, becomes primed to anticipate threats around every corner. This is why, during psychotherapy, the therapist-client relationship is so important. It is the therapist's responsibility to guide the person into balancing their stress response in the here and now, reminding them that they are safe, that the monster has gone, and that there still is a path to tread. Safety settles the brain and prepares a person for growth. Safety is the best first line of defence to threat.

Another important finding in the neurobiology of trauma is that as traumas are later relived, various regions within the **prefrontal cortex** (**PFC**—a brain region located just between your eyes, within the frontal lobes and responsible for higher-level functions such as self-awareness, informed moral judgement, the assignment of meaning to a situation, response flexibility, and self-regulation [70]) become less active.

Particularly impaired are the **dorsolateral prefrontal cortex** (**dlPFC**) and the **medial prefrontal cortex** (**mPFC**). The dlPFC is colloquially known as the brain's 'timekeeper' and is located within the PFC. It helps a person see they are in the present, not the past. [71]

The mPFC (see **Fig. 6**), on the other hand, is responsible for helping integrate the instinctive brain stem, limbic circuitry, and the higher-functioning cortex, thereby balancing the sympathetic and parasympathetic branches of the autonomic nervous system, allowing for emotional attunement and empathy, guiding intuition, and facilitating fear extinction. [72] As you can see, this is a *very* important brain area.

When the mPFC is less active, people more or less experience raw traumatic memories as

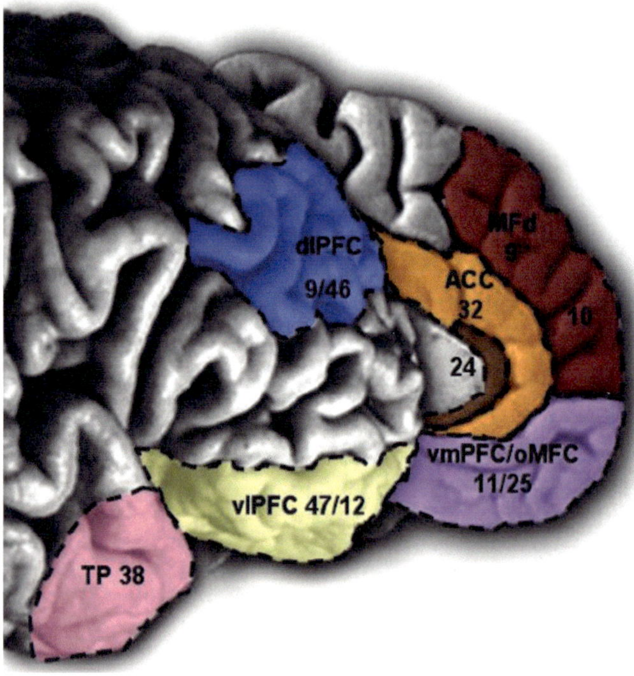

Fig. 5. Medial (i.e. middle) view of the brain from the right hemisphere (superimposed on the left hemisphere). Lateral (i.e. other side) view of the left hemisphere's PFC (in this image, the left hemisphere is partially obstructed by the right so as to show the various structures of interest). The **ACC** is the anterior cingulate cortex. The **dlPFC** is the dorsolateral PFC. The **MFd** is the medial PFC. The **oMFC** is the orbitomedial PFC (not discussed in this book, but linked to one's paying attention to internal sensations—interoceptive awareness). The **TP** is the temporal pole (located in the temporal lobe—linked with facial recognition, autobiographical memory, the naming of things, and labelling, word processing, and processing of socio-emotional information). The **vlPFC** is the ventrolateral PFC. The **vmPFC** is the ventromedial PFC.
Source: Mobbs D, Lau HC, Jones OD, Frith CD

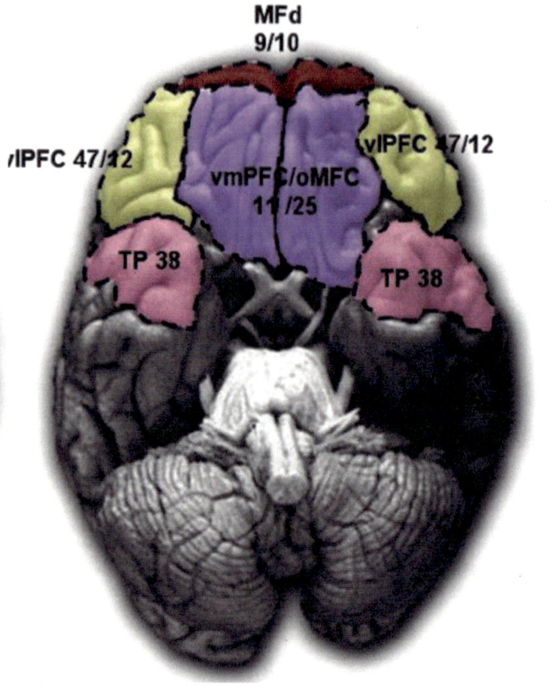

Fig. 6. From underneath: **MFd** (medial PFC—integrates brain stem, limbic circuitry, and the cortex, allowing for emotional attunement/empathy, intuition, response flexibility, and fear attenuation); **vlPFC** (ventrolateral PFC—responsible for the cognitive control required for making contextually important choices); **vmPFC** (ventromedial PFC—helps process fear and the amygdala's response to threat); **oMFC** (orbitomedial PFC—see **Fig. 5**); and **TP** (temporal pole of the temporal lobe—the latter helps process sensory, emotional, and language-based information from the environment).
Source: OpenStax & Tomáš Kebert & umimeto.org

if they are happening in the here and now, losing their ability to think clearly as they become more emotionally reactive. During this reactive state, one might find it impossible to verbalise what has happened to them. This is partly due to a significant decrease in activation within Broca's area—a part of the brain located in the left hemisphere (see Fig. 7). [73] Unable to think straight, one becomes voiceless, thrust into remembering something traumatic, and they do so in the most emotional and terrifying detail imaginable.

"He stabbed me," says Barry as he stares at me, a pleading look on his face. "The guy fuck-

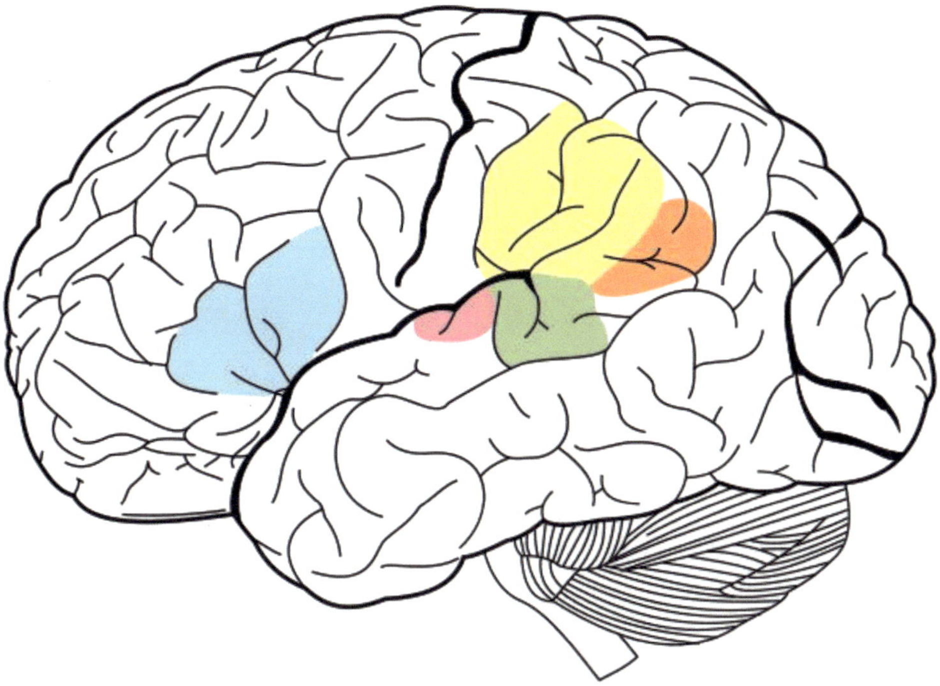

Fig. 7. From the left hemisphere: **Angular Gyrus** (transfers visual input to Wernicke's area so that we can make visual sense of what we see written down); **Supramarginal Gyrus** (processes and makes sense of language); **Broca's Area** (speech production); **Wernicke's Area** (speech comprehension); **Primary Auditory Cortex** (processes auditory input, including language)

Source: James.mcd.nz

ing stabbed me." He stands wide-eyed and lifts his shirt, pointing to his chest. I see a large purple-pink gash, still apparently healing, running from the base of his sternum to his navel. "He was supposed to be my mate." He lets his shirt drop and sits down again, wiping away a tear.

Barry remembered that night as if it were yesterday. He'd lost so much blood that the doctors weren't sure whether he'd make it. Thankfully he did, but his body still bore the marks of his assault. Although he was physically fine, his mind was still in that room. You could understand what he might have been going through. When somebody you believe is your friend slashes at you with a knife out of nowhere, there's a lot of re-negotiation happening inside. How it happened—unexpected, intensely painful, confusing—sticks with you.

From a neurobiological perspective, this is partly due to the hyper-release of adrenaline when the event was initially encoded. The experience becomes fragmented and passed into memory, except it is more vivid. The adrenal glands pump adrenaline through the

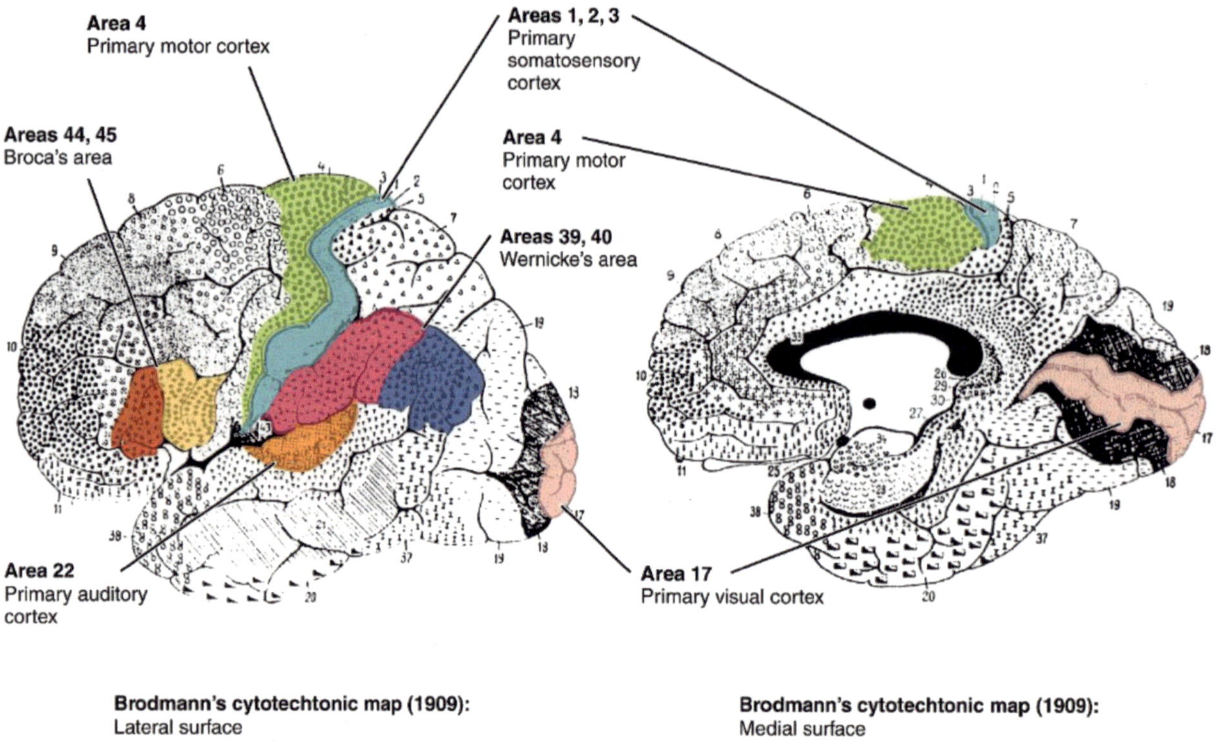

Fig. 7. From the left hemisphere, lateral and medial views of various Brodmann's areas. These brain regions were first identified by the German anatomist Kordinian Brodmann (1868-1918), showing the cytoarchitecture (Greek: *ky-tos* meaning cell; *arkhi-téktōn* meaning chief builder) of a variety of brain regions, including the **Primary Visual Cortex** and **Brodmann's area 19** beneath.
Source: OpenStax College

veins with the motive of getting one ready to tackle the threat. As this hormone courses through the bloodstream, it boosts our adaptive defences. However, the body is rarely singular in function and intention—a cascade of other impacts is observed as collateral. In particular, the hippocampi cannot process the entire event successfully, encoding only the most essential (and visceral) elements. [74] These are the most important. This often results in the brain remembering the most intense parts of the memory later. These are also often the most distressing.

At the same time, the brain experiences an increase in activation within the right **occipital lobe** (located at the back of your head), particularly **Brodmann's area 19**; see **Fig. 7**). [75] This area is associated with visual input when perception is initially experienced. [76] When something happens in your environment and your eyes pick up on the stimulus, this part of your brain lights up like a Christmas tree, rapidly encoding the experience for future recall. This is the icing on the traumatically experiential cake: high levels of adrenaline, the most physically intense element of the memory, and a clear picture to go along with it.

In psychotherapy, it is the clinician's responsibility to formulate a treatment plan to address the long-term effects of trauma on the individual. Of course, clients may know what kind of therapy they would like to receive before coming to treatment and select a clinician accordingly. However, when clients are unsure, and the therapist has insufficient skills to navigate trauma, the process becomes problematic. One example might include some person-led clinicians who follow clients where they go, rarely influencing the process (think **Rogerian psychotherapy**—the foundation, in my view, of all successful psychotherapies).

dual awareness defined

Dual awareness therapies are therapies that guide a person in keeping one foot in the present through the utilisation of various therapeutic protocols while the other foot is accessing information from the past. They promote change through the nuanced application of a variety of strategies that allow a person to consolidate the fact that they are in the here-and-now regardless of what has happened in the there-and-then. Each therapy has its own way of achieving change.

EMDR utilises bilateral stimulation; EFT and RET utilise tapping; Somatic Experiencing utilises gradual checking in with the body and enacting certain movements that have been 'stuck' during the initial trauma; focusing therapies utilise the description of a person's felt sense; Internal Family Systems therapy helps through the identification and conversation with various traumatised 'parts' of the Self.

Although dual-awareness therapies have the client's experience at their core, clinicians, too, guide the process. These therapies allow clinicians to track their client's somatic, psychological, and emotional experiences. As these experiences are labelled without being buried, the clinician then applies the relevant intervention to reduce volatility and facilitate processing.

At bottom, dual awareness therapies guide the brain into identifying, reversing, and recovering from traumatic events.

What now follows is a brief exploration into the science of how the brain changes—the science of **neuroplasticity**. During the early 1990s, brain scientists began to interest themselves more and more in the inner workings of the human brain. The years between 1990 and 1999 were termed 'the decade of the brain' due to the substantial increase in funding for research into how the brain works. During this time, research proliferated and advances were made. Scientists built upon the work of those who came before them, highlighting that the brain changes with experience. And, even more important: the brain heals.

The fact that the brain changes is now considered undeniable, as established as the finding that blood circulates rather than pools. But this fact, now taken for granted, was not always thought to be the case. For the longest time, it was thought that once the brain was damaged, this damage was immutable. [77] Little could be done to reverse these changes save invasive brain surgeries such as lobotomy or intense cranial electrification. Thankfully, we have moved beyond these methods and continue to expand our knowledge of what works for optimal human development. I find it curious to ponder what future generations will think when they look back to what we currently do, thinking it archaic and thanking their lucky stars they have moved beyond that.

In any event, we are the products of what we have available to us. We must not take the morally superior stance that the literary giant C.S. Lewis termed 'chronological snobbery', thinking that we've got it all sorted and that those who came before us were ill-informed, or worse, misinformed.

All this preamble but serves to set the stage for our discussion of how the brain changes.

It is to this that we shall now turn.

NEUR🧠PLASTICITY

the science of how the brain changes

NEUROGENESIS

the process by which new brain cells (called neurons) are formed throughout the nervous system. Neurogenesis is observed most prominently in the olfactory bulb (a brain region important for smell) and the hippo-campus (a brain region important for memory formation).

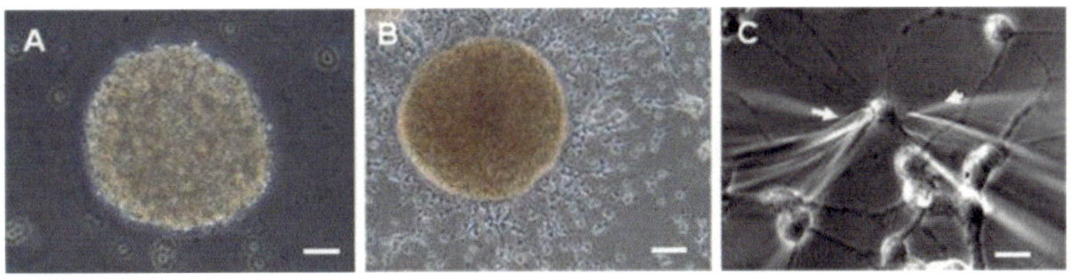

Fig. 8. Neural progenitor cells (i.e. stem cells responsible for neurogenesis) demonstrating the process of cell formation. **A.** Neurosphere (free-floating neural stem cell) after 2 days in culture (enlarged at 100 μm [i.e. micron—one millionth/m or one thousandth/mm]). **B.** Neurosphere after 4 days in culture—cells (in blue) begin to migrate from the neurosphere. **C.** Cells following formation (enlarged at 20 μm—for scale, a human hair is ~50 μm).

Source: OpenStax College

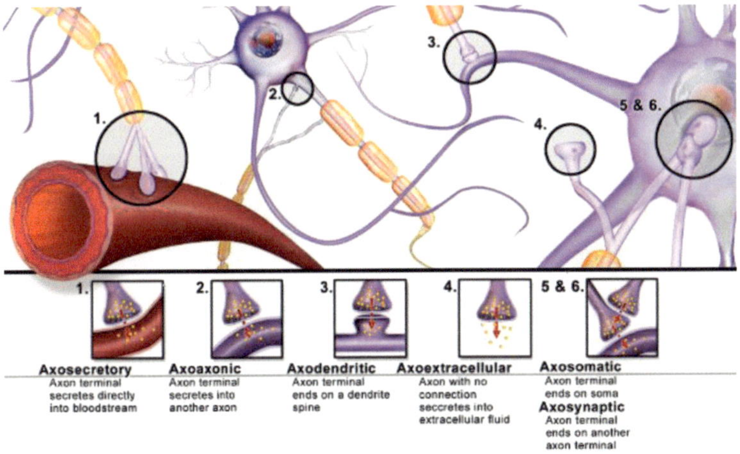

Fig. 9. This visual demonstrates various forms of synaptic communication. Synaptogenesis works to develop more of these connections throughout nerve cells, making it easier for messages to be passed along for processing. Electrochemical messages are sent through **1.** the bloodstream (*axosecretory*); **2.** onto another axon (*axoaxonic*); **3.** onto a nerve cell's dendrite (*axodendritic*); **4.** into extracellular fluid where no nerve cell lies (*axoextracellular*); **5.** onto another neuron's cell body (i.e. soma; *axosomatic*); and/or **6.** onto another neuron's cell terminal (*axosynaptic*).

Source: BruceBlaus

SYNAPTOGENESIS

the process whereby new neural connections (called synapses) are formed within the brain. As well as forming new neural connections, this process also eliminates unused synapses, replacing them with more utilised pathways, and strengthens connections that are used most.

NEUROPLASTICITY

the science of how the brain changes cont.

AXONAL SPROUTING

new axons (i.e. cell extensions conducting electrochemical messages through the neuron to the terminals in order to be sent to the next cell) form within the body of a brain cell.

DENDRITIC ACTIVATION

refers to changes at the level of the neuronal dendrite (the part of the neuron that is mostly involved in receiving information from other brain cells). Neuroplasticity has been demonstrated at the level of the dendrite (where a message is received), linking these neurons and strengthening relevant connections as they are activated.

SYNAPTIC PRUNING

the brain eliminates synapses that are no longer required—a process especially rampant as children mature and their interests become more specific, allowing for the shedding of under-utilised neural circuits. This part of the neuroplastic process is instrumental in the specialisation of a particular action or habit.

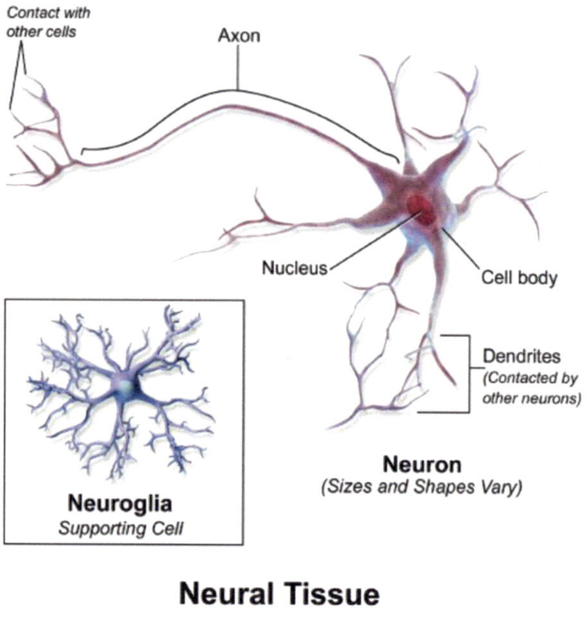

Fig. 10. This representation of a neuron demonstrates the main anatomical sites where neuroplasticity occurs. The axon terminals, though not cited, are in the vicinity of where contact is made with other cells. *Source: BruceBlaus*

Neuroplasticity is most active in areas like the **olfactory bulb** (a brain region important for smell), and the hippocampi (important for memory). The process allows for the **long-term potentiation of cells** (i.e. the strengthening of synapses for transmission of electrochemical messages). Much of what we know about neuroplasticity today was inspired by the early work of Donald Hebb [79] and Carla Shatz [80] and culminated with the Nobel Prize in Physiology or Medicine awarded to the neuroscientist and psychiatrist Eric Kandel in 2000. Kandel's work demonstrated how experiences change the neurobiological structure and function of organisms, from molluscs to mankind). [81]

2

NEUROBIOLOGY OF DUAL AWARENESS THERAPIES

You now have an introductory background on how the brain processes experiences in general and stressful experiences in particular, as well as how neuroplasticity accounts for various brain changes. From a psychoeducational standpoint, this can help you convey to your clients that deep change is possible and that there is neurological hope for healing beyond trauma.

What follows extends far beyond any one therapeutic protocol and follows that which applies to dual awareness therapies in general. I have chosen to consider these nuances from a neurobiological perspective—a popular approach that instils scientifically-grounded hope throughout the therapeutic process. This hope has come to be known as the field of **neuropsychotherapy**. It is within this backdrop that we can understand the potential inner workings of dual awareness therapies such as EMDR, EFT, RET, and many others.

"Guess the mechanics of it—why it works—I dunno. But it works; it worked for me. Seeing how this has worked... it's inspired me... If I can deal with those two issues, I must be able to deal with other issues." [82]

These words came from my research participant after I guided him through the RET protocol in an experimental case study. He wasn't bothered with the mechanics of how the protocol had worked—he was just stoked that it had. Clinically, this was significant. Theoretically, not so much. Researchers seek to understand the nitty-gritty of how a thing works.

This is what science is all about.

Therefore, I aim to present to you, the reader, six fundamental elements pregnant within many dual awareness therapies and explore their neurobiological ramifications. What follows is such an exploration.

ELEMENTS OF CHANGE IN THE NEUROBIOLOGY OF DUAL AWARENESS THERAPIES

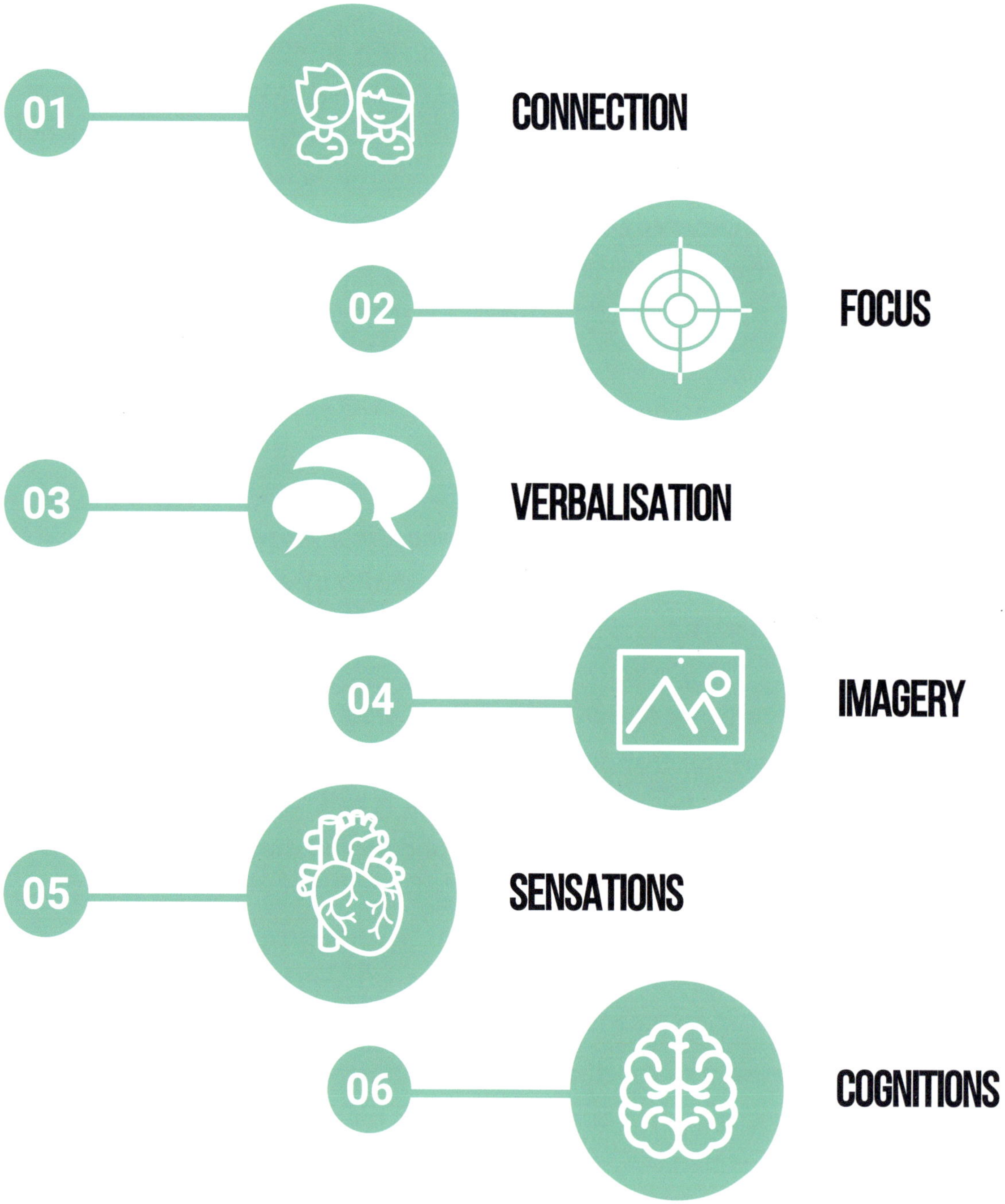

1. CONNECTION

For any psychological treatment to be effective, there must be interpersonal connection. A person coming for treatment must feel comfortable while in the room with the therapist. Therefore, it is the therapist's responsibility to create a safe and supportive atmosphere for empathic connection to develop. [83] [84] [85] [86]

Once a therapist tunes into and communicates empathically with the client, the mPFC begins its integrative work establishing a sense of safety, connection, and emotional regulation. [87] It is the work of the mPFC to focus on the safety of the here and now. As it comes online, even a very anxious person can begin to regulate.

As with any psychotherapeutic approach, dual awareness therapies require the clinician to establish an atmosphere of safety. In addition, before the processing work, many clinicians teach their clients relevant affect-regulation exercises such as the **safe place exercise** (an imagery-based mindfulness exercise that invites the client to picture a place they feel comfortable in and to notice what effect this has while mentally accessing this space) alongside **relaxed breathing** exercises. [88] Of course, there are other strategies a clinician might teach their clients, and each must be guided by their clinical competencies. In any case, a client is much more likely to engage with a novel protocol such as tapping (a protocol found in EFT, RET, and EMDR), eye movements (see EMDR), imagery-based processing work (found in Accelerated Resolution Therapy [ART]), or body-based dual-awareness protocols (such as those found in Somatic Experiencing) if there is rapport and if they have already found previous strategies beneficial before the processing work begins.

Rapport-building is an ongoing process, and once it is initially established, the client is ready to begin the information-gathering phase of treatment. This phase is found in *all* therapies regardless of orientation. A clinician must know what they are working with before administering a technique or engaging in a particular way with their client. For dual awareness therapies, the information gathered corresponds with why a client is coming to treatment. It is at the discretion of both the clinician and client to determine how much information is shared. There is no need to go into long deliberations about what happened. Communication alone is enough to build rapport, as clients often don't want to talk at length about their trauma.

2. FOCUS

For information to be gathered accurately, focus is instrumental. Dual awareness therapies utilise a client's focus by identifying a trigger to orient to while in the therapeutic space. Once a trigger is identified, the process can begin. Disinterest is the antithesis of focus and won't lead to positive clinical outcomes—if a client cannot keep track of or tune into what they need to address, therapy becomes much more difficult.

There is abundant research highlighting the various structures and functions of the human brain as they relate to focus. Any discussion herein can only be a summary. However, summarising is better than negating to mention, and it is in this spirit that we shall discuss a prefrontal region—the dlPFC—among other structures inherent in dual awareness therapies.

The dlPFC keeps track of moment-to-moment changes in experience (i.e. timekeeping), mediates working memory, and allows one to make nuanced decisions in difficult situations. It is the dlPFC that allows you to choose a particular option amongst a group of particulars. For working memory, the dlPFC helps one keep multiple things in mind simultaneously, such as remembering independent numbers (usually 7+/-2 bits of information) [89] just long enough to type them into your phone if you need to access them later. [90] [91] This process also allows for self-regulation—that is, keeping on top of one's emotions while tracking what's happening in the real world.

In addition to the dlPFC's activation in focus, this brain region teams up with the **anterior cingulate cortex** (**aCC**), a part of the limbic system that helps with emotional and cognitive processing [92]); **orbitofrontal cortex** (**oFC**) which is located in the PFC and res-

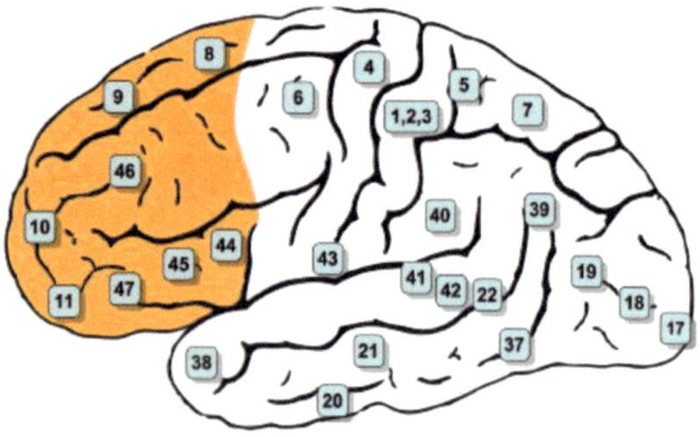

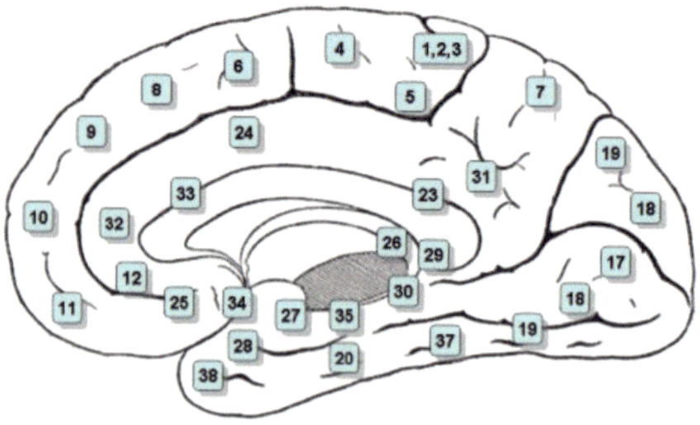

Fig. 11 & 12. These images represent various Brodmann's area (BA) locations within the lateral (see image on the left) and medial (see image on the right) cranial surfaces. The shaded area represents the prefrontal cortex. For focus, areas of import are **BA 9** (dorsolateral prefrontal cortex), **BA 24, 33, 32** (anterior cingulate cortex), **BA 11, 12** (orbitofrontal area) and **BA 44, 45, 47** (ventrolateral prefrontal cortex).

Source: Henry Gray; Bartleby.com

ponsible for functions like visual recognition and emotional reward/punishment; [93] and the **ventrolateral prefrontal cortex** (**vlPFC**), responsible for the cognitive control required to make contextually important choices. [94] As these areas activate, the PFC assesses information from the external world and responds effectively. [95]

As traumatic memories flood the senses, the brain loses itself and confuses what's being recalled with what's happening in reality, losing the sense that these memories are history. Dual awareness therapies may help serve as a conduit for reclaiming the present when experiences are later recalled. This recall allows the dlPFC to do its work in organising experiences while noticing associated images and emotions (the territory of the oFC), sensations (arising from the body proper), and thoughts (mediated and regulated through the vlPFC). All this helps one remind oneself that the memories are coming from a dead past and present no actual threat in the here and now.

By utilising the strengths of various prefrontal and limbic brain regions, the therapist helps the client reclaim so-called 'cortical real estate', [96] reinstating a lost sense of control to the person with each focus of the mind. On top of this, the mPFC—a brain region whereby all these cortical structures converge—signals the release of gamma amino butyric acid (GABA for short) to calm the brain and lower the release of the stress hormone cortisol—a finding that represents numerous studies using dual-attention techniques such as EFT and EMDR in their methodologies. [97] [98] This process, by extension, teaches the brain there is nothing to fear right now. [99]

3. VERBALISATION

During the information-gathering phase, dual awareness protocols utilise:

1) verbal acquisition of data (done in tandem with the client); and
2) verbal repetition of acquired data.

This means that the person is not a silent rider—they are intricately involved in communicating what is going on for them throughout the entire process. But why is this important?

When a traumatised person re-experiences a triggering memory, one of the first things the brain does is quiet the left hemisphere's Broca's area. If you recall, this is the brain region linked to speech-production. The ability to talk, to verbalise what's going on for them, is significantly impeded, and the person finds it difficult to get their point across.

"I just can't say the word," said one of my clients who had been sexually assaulted. For her, the word *rape* was impossible to say at the time. Her words failed her.

Martin Scorsese's *The Irishman* [100] provides an example of this shutdown in communication. In the film, one of the protagonists, Frank Sheeran, has just murdered his good friend Jimmy Hoffa. Shortly after, he receives a phone call from Hoffa's wife. Sheeran stammers, stumbles over his words, cannot connect emotionally with her, and cannot get his words out.

In a safe and supportive environment, dual awareness therapies help clients verbalise elements of an experience with the aim of resolution. These therapies help individuals keep important information at the front of their minds by guiding focus towards some of the main elements of the memory. For tapping therapies, this is done, for example, by reciting a variation of the following statement: *Even though I feel* [list emotion and/or bodily sensation] *when I see* [list a visual of the traumatic memory] *I am here now. I can [learn to] accept myself.* The way this might look like in therapy is as follows: *Even though I feel my heart racing and a lump in my throat when I see that car racing coming at me, I'm here now. I can accept myself.*

The above statement is but one adaptation of what you'd find in tapping therapies, and it is not found in therapies like EMDR. In EMDR, bilateral stimulation occurs silently as the client is guided to reflect on and remember what they would like to feel differently about. It is in the time preceding and following the administration of the protocol that verbalisation occurs. In therapies like Somatic Experiencing or Internal Family Systems therapy (whereby a person is guided in visually accessing and integrating varying 'parts' of oneself while remaining in the present and discussing with the therapist what comes up), verbalisation occurs throughout the process. Of course, each therapy has its nuanced approach, but speaking about what comes up is often pivotal.

Once a person has highlighted the stressful elements of the experience while being guided through a dual awareness protocol, there is often a reduction in overall **subjective units of distress** (**SUD**) as scored on a 0-10 SUD scale. [101]

It is theoretically possible that the act of orienting to the present—a function of the dlPFC—and verbalising elements of what had happened—a function of the left hemisphere's Broca's area—balances the emotionally-dominant right hemisphere. As the right hemisphere (a chunk of the brain that is hyperaware of emotional and traumatic experiences) reduces its laden affect—and the client's focus is on the clinician—there is a higher likelihood that the trauma can become integrated. [102] The brain begins to recognise that the event is not happening in the here and now, that it is no longer threatening, and that connection is possible without the threat of being judged.

4. IMAGERY

Many therapeutic interventions use imagery to guide a person towards change. Visual recall is paramount whether it's bringing up a dream in treatment, asking a person to relay a memory, or exploring how life looks like a few years into the future. Some therapies focus primarily on visualisation, excluding things like cognitive strategies. ART, for example, guides a person into visualising an alternate ending to a traumatic incident, facilitating visual mastery over the seemingly inescapable situation. [103]

When my eldest son was about eight, I helped him process fears he had about going to bed. At the time, he was afraid that Pennywise the clown might come out of nowhere and obliterate him. Employing elements of ART combined with Imagery Rehearsal Therapy, [104] I guided him to imagine the picture stuck in his head—it was an evil clown grinning down at him, blood trickling down its fangs. Then, I guided him in imagining something he'd like to do to that clown that makes him (my son) feel more in control. He pictured Pennywise running around in circles, scared, farting as he went. Needless to say, this led to him laughing his way out of that fear.

From a neurobiological perspective, visualisation approaches make sense in that they take an uncontrollable situation and provide a person with a way of taking back control, if only visually. Recent neurocognitive research has demonstrated that the following brain regions are active in this process: 1) the hippocampi; 2) the temporal lobes' auditory cortices (brain areas that process sound); 3) the **ventromedial prefrontal cortex** (**vmPFC**)

—a brain area that regulates emotional decision-making; and 4) the **nucleus accumbens** (located within the **mesolimbic pathway**, which connects the midbrain to the limbic system) which help to mediate desire and reward through the release of dopamine, facilitating the anticipation and reinforcement of motivating experiences. [105]

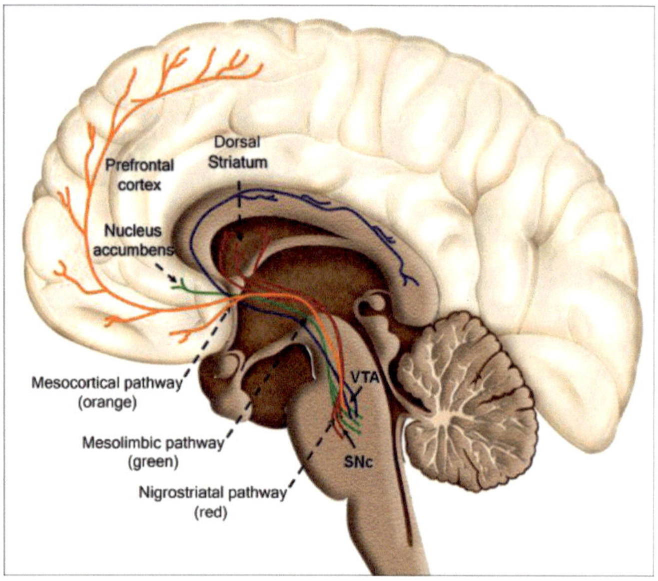

Fig. 13. The green line represents the mesolimbic pathway, containing the nucleus accumbens at the top branch. Importantly, activation in this area represents a reinforcement of a particular motivating behaviour.
Source: Arias-Carrión, Stamelou, Murillo-Rodríguez, Menéndez-González & Pöppel

These cranial areas are activated upon recall regardless of whether the images are real or imagined. [106] [107] [108] More important, as images are imaginatively re-experienced in a safe environment, the nucleus accumbens activate, playing a significant role in decreasing associated fear and instilling a sense of mastery in the individual. [109]

Just by bringing these images up—staring the dragon in the face—the fear begins to fade.

The importance of these findings for dual awareness therapies is compelling. With imaginal triggers, one can no longer say that it's only actual experiences that carry a neurobiological punch. If thinking about an event can cause the body to go into overdrive, we need a strategy to address this. Fear associated with an image can often be extinguished or powerfully attenuated through recall and adjustment in a safe environment. This makes practical sense because if one feels competent, trusting their own ability, they are more likely to take that cautious step forward. As the memory's hold is released, the brain's reward circuitry gets into full swing, and the action is reinforced—one feels more in control, courageous, and powerful when taking the devil by the horns.

This bears out in clinical practice. Emerging research has shown that traumatic images reduce in intensity and even change altogether during and following dual-awareness protocols, including the RET protocol. [110] I once had a client who transitioned from one image of his abusive father-in-law grabbing him by the shoulders and violently shaking him while they were at a footy game—my client was nine at the time—to playing in a park with his sisters, smiles on their faces a moment later. This transition occurred immediately following my administration of the RET protocol.

It was a compelling experience.

5. SENSATIONS

Traumatic re-experiencing is often as visceral as would be if one were plunged into a frozen lake—it's a whole-body affair that occurs within the viscera. Even before traumatic stress was codified as PTSD in the DSM-III, several body-based therapies, not least Somatic Experiencing, were being utilised to treat trauma. [111] [112]

With the uptake of neuroscience in trauma treatment, the body and the brain's experience have become central. One of the most useful neurobiological frameworks for understanding traumatic stress is Stephen Porges's **Polyvagal Theory**. [113] In part, Polyvagal Theory proposes that the 10th cranial nerve (the vagus nerve) regulates many major body systems, such as the cardiovascular, gastrointestinal, and sympathetic and parasympathetic nervous systems, ultimately keeping them in check. [114]

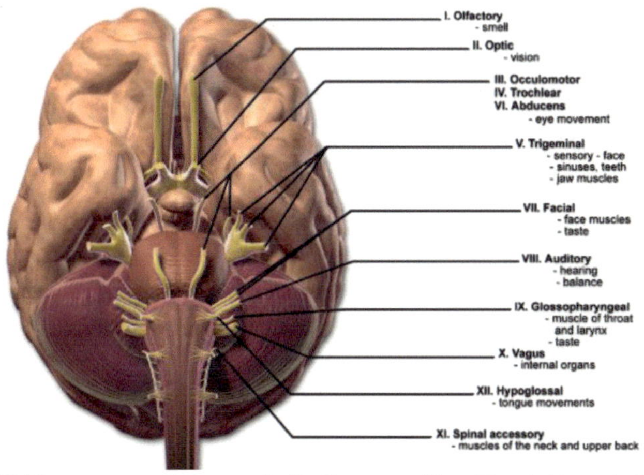

Fig. 14. The vagus nerve is represented by the numeral X. and is noted to control for many internal organs. The sympathetic and parasympathetic nervous systems are represented on the next page.
Source: BruceBlaus

Porges has written extensively on the processes by which this occurs, but it is essential to understand that the more one knows about how the body responds to an external stressor in a given situation, the more that person can select a necessary corrective response. [115] If one knows what to expect, resolution is closer at hand. By way of example, years ago I was speaking with a Somatic Experiencing practitioner who I later interviewed as part of *The Emil Barna Podcast* Robert Watson. Watson taught me that

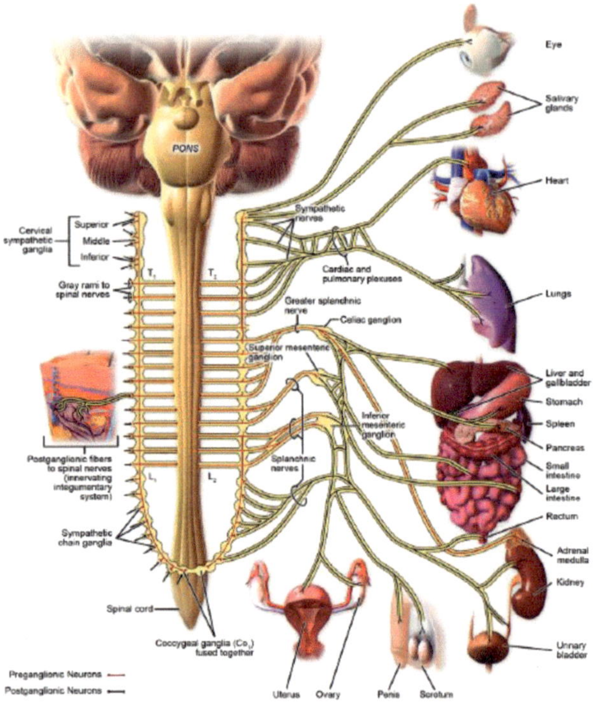

Sympathetic Innervation

Fig. 15. As noted in **Fig. 14.** the vagus nerve controls for many internal organs, including many of those that innervate (i.e. supply nervous influence or control, stimulating aspects of the required system through cranial or peripheral nerves) the sympathetic nervous system. The sympathetic nervous system gets the body ready for action.
Source: BruceBlaus

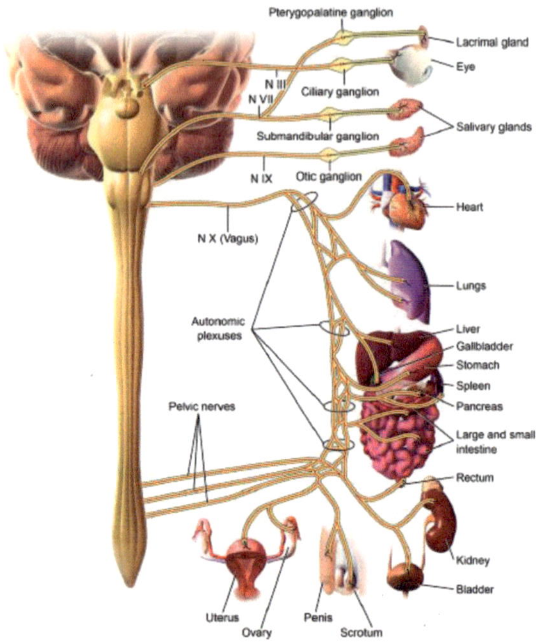

Parasympathetic Innervation

Fig. 16. Just as the sympathetic nervous system excites the body into action, the parasympathetic nervous system (the second branch of the autonomic nervous system) slows things down. Like a parachute, it calms the human organism by working towards homeostasis. Clinically, this can be demonstrated when a person feels more relaxed, their heart slows, breathing becomes deeper and they might even yawn or swallow.
Source: BruceBlaus

the fight response occurs above the waist (e.g. tense shoulders and face, clenched jaw and fists, racing heart, butterflies in the stomach) while the flight response occurs below the waist (e.g. restless legs, the urge to run, tense quadriceps and calves). The freeze response, on the other hand, flows throughout the body and appears as a sort of shock to the system; while the fall/feign responses appear like a blunting of experience (e.g. drooping shoulders, lethargy, downward gaze, zoning out, dissociation).

In the face of traumatic distress, each of these responses represents the opportunity to utilise a particular strategy to get the system back online. This is what Somatic Experiencing does so well, but also what other dual awareness therapies emulate by bringing the person back into the present instead of leaving them locked in an unchanging past.

Tapping protocols, EMDR, ART, and even mindfulness practices conceptualise information relating to bodily sensations and paired emotions as integral components of implicit, procedural (i.e. body-based) memories. Clinicians focus on what is happening in the client's body and how they are experiencing it.

To linger on this point for a moment, one must consider procedural memories. These memories are those which require no thought to complete. Examples include riding a bike, brushing one's teeth, walking, running, grappling, balance, and so on. These actions are so etched in our way of being that they become subconscious. Their habitat? Areas within the brain such as the **cerebellum** (located behind the brainstem and associated with movement and balance), the limbic system (responsible for stress perception, memory, and other instinctive drives), and the **extrapyramidal system** (see **Fig. 18.**). [116] [117]

In any situation, there is a delicate interplay between physiological arousal and focused calm. With traumatic distress this dance is interfered with

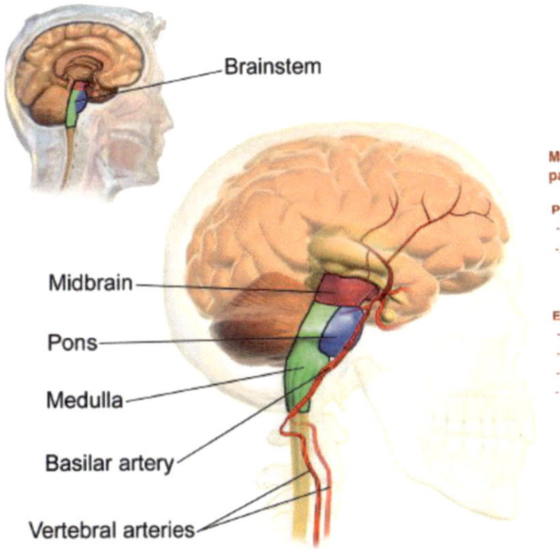

Fig. 17. A visual representation of various structures of the brainstem. The dark brown structure (unlabelled) towards the bottom-left of the brain is the cerebellum (controlling for balance and movement). The medulla (pictured in green) is where the extrapyramidal system is located (see below).
Source: BruceBlaus

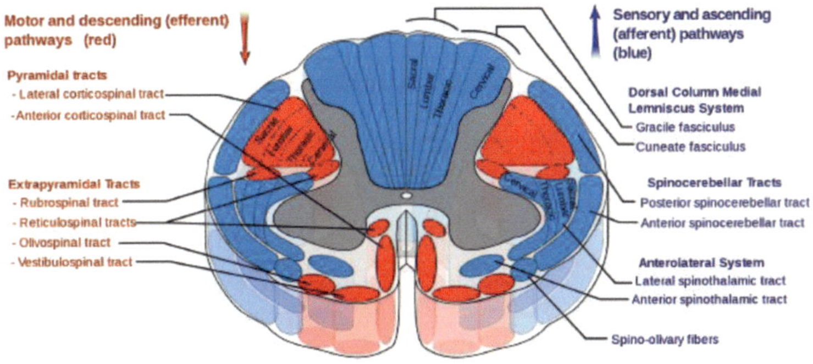

Fig. 18. The extrapyramidal system controls for involuntary movements and is termed extrapyramidal as it is located outside the pyramidal tracts within the medulla (pictured above in red as triangular shapes). This system is made up of motor neurons responsible for reflexes, posture, complex movements that were once voluntary but have become 'muscle memory' (e.g. dancing, grappling, writing, etc.), and locomotion (ongoing movements taking one from one place to another).
Source: BruceBlaus

and can be skewed into favouring physiological arousal at the expense of homeostatic calibration. In other words, a person becomes highly wired with significant baseline anxiety levels. They find it difficult to calm down, and even the concept of calm can become triggering as hypervigilance pushes them to always look for what's around the corner.

Dual awareness therapies aim to bring this system back into equilibrium. 'Renegotiating' [118]—a Somatic Experiencing term that describes the process of exposing oneself through various means to a bodily reaction, slowing the reaction down, and pairing it with other bodily actions to help 'complete' what the reflex set out to do in the first place—helps a person turn reflexes into mindful appreciation and an overcoming of what is going on in order to give one more control over their nervous system. The process of 'renegotiation' is inherent in any successful therapy, including tapping and EMDR. In the case of tapping, the identification of somatic experiences paired with the tapping protocol often creates a sense of mastery over once uncontrollable sensations. Importantly, mastery is tantamount to skill, and skill builds courage and confidence.

What is neurobiologically fascinating is that the process of *pushing through* despite discomfort while remaining present and mindful activates a part of the limbic system called the **anterior midcingulate cortex** (**amCC**). [119] The amCC is associated with dopamine and noradrenaline release, motivational drive, and reward-oriented behaviour. [120]

Pushing through despite your feelings about it increases your motivational drive and, incidentally, associated reward. The neuroscientist Andrew Huberman, PhD, calls this push in spite of adversity *limbic friction*. One part of your brain—your limbic region—is telling you to *Get the hell out of here!* while others—namely, the prefrontal regions of your cortex—are coaching you through it. It is often in the pushing through of this 'limbic friction' that one grows. Clinical depression stunts this ability, which is why so many behavioural (and cognitive) strategies are provided in the treatment thereof. However, for those with posttraumatic stress, the navigation of said friction can be vital for healing. Remember, however, that it is not only in the daily grind that growth is recognised but in the appreciation and acknowledgement of a sense of mastery, no matter how small. As the legendary phil-

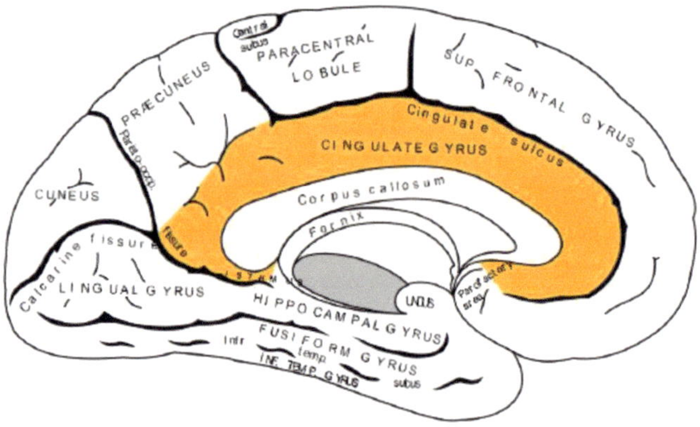

Fig. 19. The yellow shading represents the cingulate cortex in the image above. The anterior cingulate is located at the front, between the *fissure* and the *isthmus* (Greek for a 'narrow area that connects two larger areas'—in this case, the corpus callosum and the hippocampal gyrus). The amCC is the anatomically medial area of the aCC.
Source: Henry Gray

osopher and Brazilian Jiu Jitsu coach John Danaher said in an interview with the MIT scientist Lex Fridman: "Confidence is *always* paired with physical ability in pressing situations, bar none." (paraphrase)

Returning to dual awareness therapies, through somatic recall (bringing up a feeling) and attenuation thereof (noticing a shift following an action), one becomes motivated to continue working through it. Clinically, I see this when a person begins to frown as they come to terms with their changing mood and bodily sensations. Entering the session, they might have believed they would always be stuck in their reactions. Upon leaving, they have a different experience. This experience becomes a beacon of hope.

The amCC is tightly linked with the **insula**—a small cortical region responsible for social, emotional and cognitive processing—and visceral functions; [121] and communicates with the amygdala, hypothalamus, thalamus, and brainstem in calming the brain and allowing for the rescripting of fearful somatic reactivity. [122] In effect, the amCC is one of the only cortical regions that helps dampen neural firing in the amygdala associated with fear formation, thus reducing perceived distress. [123]

One might infer that as mindful attention is directed to somatic sensations—while paired, in the moment, with dual awareness techniques—activation within the amCC works alongside other cortical structures to attenuate the amygdala's fear response. This proposed reduction can, in turn, lead to parasympathetic activation. As the parasympathetic nervous system comes online, the experience is renegotiated, and instinctual procedural memories become less hostile. Clinically, this creates a sense of mastery, especially when revisiting the memory later.

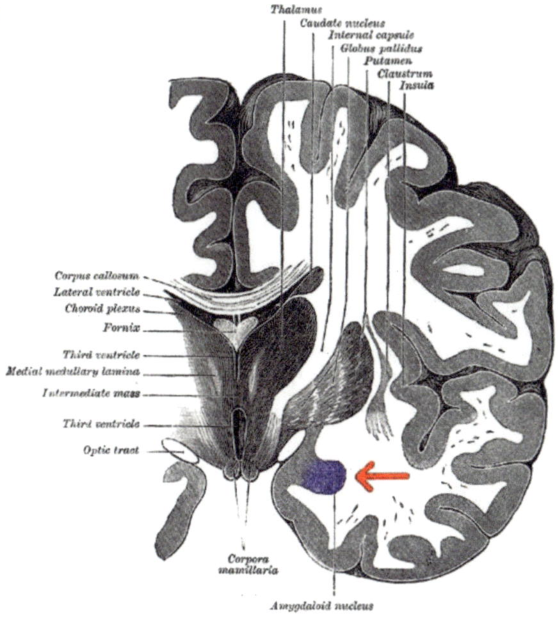

Fig. 20. The amygdala is represented by the purple shading in the image above. This area has been observed to increase in size with chronic stress and decrease in size once stress is attenuated over some time. There are two amygdalae located in the human brain: one in the right hemisphere (above) and the other in the left.
Source: Henry Gray; Bartleby.com

6. COGNITIONS

Our final stop in conceptualising traumatic memories as they relate to dual awareness therapies rests upon a person's mental life. Stressful and traumatic experiences can change the way one sees the world. They can change the very core of the beliefs one has established about oneself, others, and the world writ large. These beliefs often lie under conscious awareness and are termed **schemas**.

Schemas can be conceptualised as implicit **engrams** (i.e. cognitive information imprinted upon a physical substrate—the brain—informed by the experiences that shape certain types of memories and ways individuals see themselves). These engrams are proposed to change the physical structure of the human brain and associated neural networks. [124] [125]

Experiences conjure up repetitive thoughts such as *I'm not safe*, *People can't be trusted*, *I've gotta keep my head down and guard up*. These thoughts can lead to the development of certain schemas such as Mistrust/Abuse, Self-Punitiveness, Unrelenting Standards, Self-Sacrifice, Subjugation, and so on. [126] As schemas form, representative neural pathways may become activated and associated with these thoughts, feelings, and fears.

In EMDR, the AIP model speculates that as noxious psychological stimuli are re-activated, there is a block in neural processing and the EMDR protocol helps to rectify this block. [127] This conceptualisation might also apply to other dual awareness therapies like tapping.

As we have already seen, the neurobiological processes at work in the activation of traumatic memory and the processing thereof are multifactorial. As clinicians utilise dual awareness therapies through various protocols, thus navigating the neural terrain of trauma, they first and foremost help the treatment-seeker access the memory. This activates associated neural substrates which include those accounting for imagination, bodily sensations, focal awareness, verbalisation, and schematic/cognitive information. Throughout this process, there is the opportunity for one to address thoughts associated with the trauma.

Some research has found that as memories are accessed, there is a short-lived time window in which they can be reconfigured. [128] Each time a memory is accessed, that memory is malleable to change. As one neuroscientist puts it: *Memory retrieval is a memory modifier.* [129] [130]

Many dual awareness therapies utilise protocols to minimise a person's SUD scores. Once SUD decrease to ~0/10, the client is guided to revisit the initial thought(s) attributed to the memory and rate their emotional valence before revising their initial, trauma-laden thoughts in a new light. Clinicians help the person come up with alternate cognitions that serve to challenge the negative. Oftentimes, the positive cognition (e.g. *I am strong*) is in direct opposition to the initially encoded counterfactual (e.g. *I am weak*).

Some therapies rate the validity of the positive thought as it relates to the initial negative thought (think EMDR); others seek to 'encode' alternative thoughts following processing (consider EFT and RET). For therapies like Somatic Experiencing, there is no explicit encoding process, but more of a bottom-up focused discussion of how a thought, feeling, emotion, and/or sensation affects the person as the body renegotiates its relationship with the traumatic experience. As you can see, each therapy has its approach.

From a tapping perspective, once the positive cognition is accessed and related to the initial memory through the tapping protocol. This is done with the clinician incorporating a statement such as *Even though this memory might come up again, I choose to know that I am* [insert positive cognition] *now.* This statement is simultaneously paired with the EFT tapping procedure, leading to a decrease in SUD. From a brain-based perspective, this process can activate other neural circuitry to contend with the original neural map. The difference now is that this new, more adaptive set of cognitions associated with the original memory. Repeated often enough and paired with positive visceral responses (e.g. a calmer affect), a new and more adaptive experience can take hold.

This brings us full circle to Hebb's Law: *Neuron's that fire together, wire together.* Now, it is the more positive and uplifting thought that takes precedence over the older, degenerative one.

3

SYNTHESIS & CONCLUSION

And now, we bring home the threads of discussion to drive the points argued for throughout this book: that the neurobiology of trauma is multifaceted and, in turn, hope for effective treatments is vast.

I have argued that dual awareness therapies demonstrably overlap and ought to be conceptualised as such. In the same vein, I have argued for some nuances, especially from my vantage point, having been trained in RET but having seen the utility of other approaches in allowing for effective outcomes.

Dual awareness therapies are integrative and holistic psychotherapies that consider neurobiological systems, cognitions, emotions, visual representations, and somatic sensations to address recurring stresses and traumas. By utilising current neurobiological findings, I proposed a (tentative) understanding of how these therapies might work.

From the hub of the mPFC—which creates a sense of connection and provides a safe launching pad to mitigate the fear response—the clinician empathically resonates with the client, providing a safe space to address what's brought to therapy. As rapport is established, a treatment target is identified, and the relevant protocol is engaged. As a relevant dual awareness therapy is utilised, various prefrontal regions of the brain are active in bringing the person into the present—the here-and-now is utilised to address the there-and-then. The left hemisphere of the brain takes a more active role in recalling elements (e.g. words, thoughts, images, feelings) salient to the initially encoded experience(s).

Visualisation and mindful acceptance of somatic sensations are then brought to the fore. In doing so, a sense of mastery is achieved once stressful images and sensations melt away. The brain encodes more positive visual alternatives, and a person's physical state relaxes. Prefrontal cortical regions work in tandem with the autonomic nervous system's sympathetic and parasympathetic nervous branches. Various limbic regions are brought under control. A time-and-place stamp is placed on the memory as the hippocampus comes online, and the target loses its negative valence. The person reorients him- or herself to a feeling-state much better than while in the throes of trauma, and the memory no longer packs the same emotional punch as before. The person feels more in control. In short, the memory is reprocessed.

I expect that this book and the neurobiological correlates explored are not without their challenges. These ideas are in some ways speculative, drawing upon extant neuroscientific literature that has not yet been applied to some of the protocols described herein—for example, the RET protocol. There may also be other mechanisms in operation that I don't know of yet, therefore, have not included. Nevertheless, it has been my practice to represent as much factual data as I know how to, not being a neuroscientist but a clinician, and describe this data in a clinical light. If this book stands up to scrutiny, the proposals found therein may become generalisable to dual awareness therapies at large. After all, as I have sought to make clear, the mechanisms at play are relevant not only to one therapy or another but to *any* therapy utilising similar means of intervention.

And it is on this note that I shall end. Thank you, dear reader, for taking the time to explore with me.

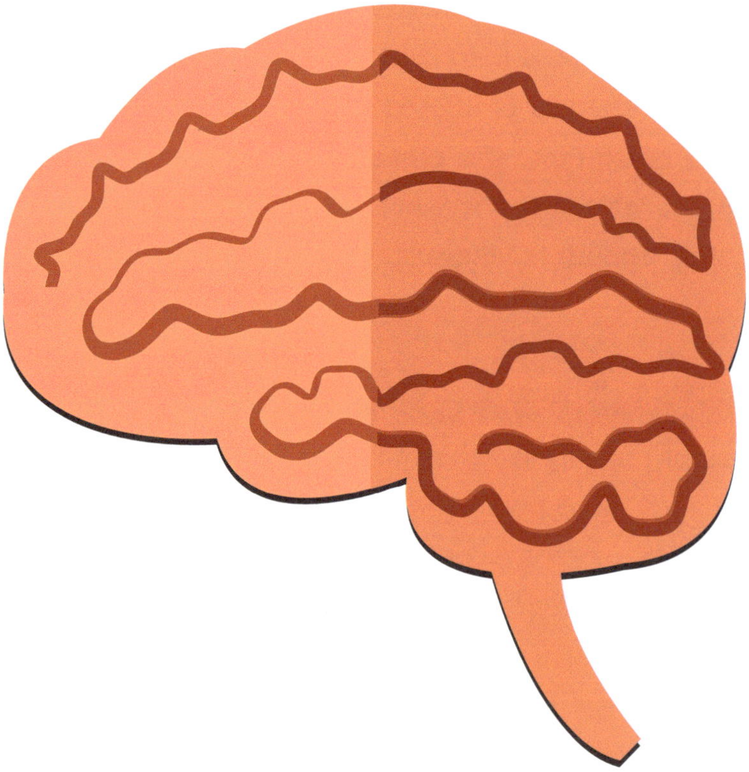

GLOSSARY & COMMENTARY

aCC: see **anterior cingulate cortex**

acupuncture: a therapy that involves the insertion of very thin needles into the skin's surface (between 3mm to 2cm, depending on the area of the body) to help alleviate stress and physical pain. Inserting these needles can release endorphins which some believe is attributable to pain reduction.

adaptive information processing model: a conceptual model explaining the outcomes achieved by EMDR. This model proposes that stressful memories are stored in an unprocessed, affect-laden manner and can result in blocked neuronal processing. Noxious psychological stimuli are later re-activated at a neural level—somatically, cognitively, imaginally—and can leave an imprint that is difficult to budge. Some theorists posit that EMDR helps rectify this block by revisiting traumatic memories in a safe environment while pairing them with bilateral stimulation.

adrenal glands: located atop the kidneys and, in the context of stress, influenced by the pituitary gland (which secretes the adrenocorticotropic hormone) to further secrete cortisol and adrenaline in order to prepare the body for action. The adrenals operate in the context of the hypothalamic-pituitary-adrenal (HPA) axis to signal the appropriate release of chemicals required for the stress response.

adrenaline: a hormone that is produced by the adrenal glands to motivate the body towards action. Some individuals report feeling the sensation of adrenaline just beneath their diaphragm, in the middle of their gut, when they feel under pressure. The felt sense is often described as a wave of energy that is paired with a person's holding their breath and a sharper focus. A sense of dread, however, can also occur under certain circumstances. Adrenaline is also referred to as epinephrine (a neurochemical in the brain).

affect regulation: the regulation of a person's observed emotional state. For example, a person can be described as having a despondent affect by an observer (e.g. withdrawn posture, drooping shoulders, and a downcast, sombre expression). Affect is different to mood, which is a person's *self-report* of their emotional state. Affect regulation can be described in the context of non-psychotherapeutic and psychotherapeutic strategies which help neutralise a negative emotional state and/or build towards a positive mood. Some examples include listening to music, watching a movie, laughing, smiling, engaging in breathing exercises, exercise, prayer, etc.

AIP: see **adaptive information processing model**

akathisia: a severe adverse reaction to psychiatric medications that can lead to involuntary muscular contractions and pain. For some, akathisia leads to suicide. One client of mine described her reactions to the antipsychotic medications Quetiapine and Risperidone as so unbearable that when the psychiatrist told her he would be increasing her dose, she had a full-blown psychotic episode, fleeing the psychiatrist's office, speeding home, and physically assaulting her father (who had severely traumatised her while she was a child) with a knife. Common symptoms of akathisia include severely restless legs, arms, and body; feeling extraordinarily lethargic and sedated but unable to stop moving.

alternative therapies: therapies that lie outside of the medical model of treatment. Examples include 'energy psychology' techniques (e.g. EFT), chiropractic care, dietary therapy, herbal medicine, mindfulness/meditative practices, acupuncture, aromatherapy, and yoga. These therapies are becoming more mainstream as research begins to accumulate.

amCC: see **anterior midcingulate cortex**

amygdala: a pair of neural clusters located within the limbic region of the brain and responsible for emotional activation, fear, memory, and attention. These neural structures grow in mass and reactivity with chronic stress and can influence a person's stress response, making them hypervigilant as time goes by.

amygdalae: see **amygdala**

anterior cingulate cortex: a neural region that is nestled between the limbic system and the prefrontal cortex, sitting atop and wrapping around the corpus callosum. It is located towards the front (anterior means 'close to the front') of the cingulate cortex, linking the cortex to the limbic system and responsible for emotional and cognitive processing, ethical decision-making, self-control, and self-regulation. The anterior cingulate has a unique function, helping you think about the types of emotions you're experiencing and deciding whether it's best to react or respond.

anterior midcingulate cortex: a part of the anterior cingulate cortex associated with movement, especially intentional motor movement—that is, when we choose to move in a particular way. Cortical activation in this area is often associated with dopamine and noradrenaline release, motivation, and reward-oriented behaviour. Practically, when you overcome an obstacle (physical or mental), this area lights up and reinforces your sense of accomplishment.

arrest and orient: a universal response to threat detection. An organism—animal or human—stops to assess their environment, scanning for potential danger. This happens in day-to-day life whenever one needs to collect oneself and look around their immediate vicinity to see what's up.

autonomic nervous system: the body system involved in maintaining automatic bodily functions. This system operates without conscious awareness, regulating heart rate, temperature, respiration, perspiration, arousal, etc. One way to see this system in action is to focus on a specific point outside the window when you're on a train or another moving vehicle. You'll notice your pupils track the moving object until it is out of sight before flicking back to the middle field of vision.

axonal sprouting: the process whereby new axons (neuronal extensions protruding from the cell body) form within the body of a brain cell. Axons carry electrochemical impulses from the top of the neuron to the terminals down the bottom. Depending on the axon's insulation (i.e. myelin sheath—a fatty kind of axonal covering that helps contain and supercharge electrical impulses, like how an electrical cord wraps around a wire to insulate and contain a charge), the resultant message is sent at high or low speeds.

behaviourism: the paradigm by which one understands that every behaviour—positive or negative—is learned through conditioning. The more something is done (say, the bell at school signalling home time), the more likely that thing will be conditioned into behaviour (e.g. preparing to leave for home).

behaviourist tradition: Behaviorism originated in the U.S. with B. F. Skinner (1904-1990) and demonstrated (through experimentation) how certain behaviours achieve particular results. This tradition underpins many cognitive and behavioural treatments and can be very useful when developing new routines and pairing various actions to create a more disciplined lifestyle.

bilateral hemispheric stimulation: see **bilateral stimulation**

bilateral stimulation: stimulation of both hemispheres of the brain. In therapy, this can be accomplished through eye movements (e.g. looking from side to side while following the therapist's fingers) and switching between emotional and verbal content (e.g. discussing feelings and then switching discussion to mathematics immediately afterwards).

biological paradigm: an often deterministic understanding that holds human activity and behaviour as determined by one's biological makeup (i.e. genetic predisposition, primary sex characteristics, temperament, physical capability, etc.). This paradigm is instrumental in psychotherapy to educate clients about how their body responds to various situations. However, it is not helpful if these explanations are used as reasons to justify one's behaviour. For example, this is often seen in groups like Alcoholics Anonymous, where addiction is seen as a brain disease. In my clinical experience, this leads to a person taking on an 'addict identity' where they only define themselves by their biology.

body system: (also referred to as an organ system). There are 11 systems within the human body that perform their own unique functions. These are the: ***nervous system**; **endocrine system**; **reproductive system**; **sensory system**; **cardiovascular system**; **respiratory system**; **immune system**; **digestive system**; **muscular system**; **skeletal system***; and the ***integumentary system***. Each works in tandem with the other to help the body along in its day-to-day functioning.

brainstem: a vital brain structure located atop the spinal cord but below where the 'actual brain' begins. It serves to connect the cerebrum (also called the cortex: the largest part of the brain) with the spinal cord. The brainstem is responsible for the body functions that keep us alive, such as breathing, heart rate, and communication between the body proper and the brain through the cranial nerves. This is partly the reason why with severe head trauma one can die instantaneously. Compare this with physical trauma to the body —death may take much longer.

Broca's area: a small cortical region in the left hemisphere (located in the posterior frontal lobe) and responsible for speech production. If a person suffers damage to this part of the brain, they cannot speak. It is located close to another area responsible for speech comprehension: Wernicke's area. People who are stressed find it hard to communicate when they are triggered.

Brodmann's area 19: an area in the right occipital lobe (towards the back of the brain) that registers visual input when perception is initially experienced. This part of the brain allows one to perceive visual stimuli (i.e. images, shapes, and features) and integrate them into consciousness, thereby helping one make sense of the whole experience.

CAT scan: computerised axial tomography. An imaging technique that takes X-rays of various body systems and reproduces a static, black-and-white picture. It is used to analyse body structures rather than functions. Specialists use these types of imaging techniques to see if there are any abnormalities within the human body.

central nervous system: consists of the brain and spinal cord.

cerebellum: the sponge-looking part of the lower brain—the hindbrain—located in the area of the brainstem. This part of the brain is responsible for movement and helps you balance and posture yourself. It can be that part of you that helps you present confidently, move well, and feel in control.

cognitive revolution: a revolution that challenged the prevailing behaviourist paradigm of the time. Many cognitive therapies came out of a shift from behaviourism to a cognitive understanding of human behaviour. In particular, CBT.

cortex: in Latin, cortex means 'bark' as in the bark of a tree. It is the brain's outer layer, consisting of about 28 billion neurons. It is most developed in human beings and higher primates (i.e. chimps, bonobos, and orangutans). It is also responsible for higher-level thinking and processing.

cortisol: a stress hormone produced by the adrenal glands, released when you awaken, are feeling stressed, or are getting ready for action.

dendritic activation: refers to changes in the area of the dendrite (i.e. the part of the neuron that is mostly involved in receiving information from other brain cells). Neuroplasticity has been demonstrated at the dendrite level (where electrochemical messages are often received)—a process whereby neuronal connections are strengthened as they become activated over time. Importantly, the term 'dendritic activation' does not appear in medical sources, but I thought it fitting to use it in describing the action of dendrites during neuroplasticity.

depersonalisation: an uncommon experience in the clinical sphere where a person forgets and/or feels separate from oneself. One may forget who they are, forget core aspects of themselves (i.e. what makes them *them*) and, in extreme cases, experience personality splitting such as is observed in Dissociative Identity Disorder (previously known as Multiple Personality Disorder).

Diagnostic and Statistical Manual of Mental Disorders: the psychiatrist's Bible. It was first published in the early 1950s and is currently in its fifth—and largest—edition yet. This edition was revised in 2022 and, at the time of this writing, is known as the DSM-5-TR. The DSM's nomenclature is utilised worldwide—though more prevalent in western countries—in order to describe and categorise mental health conditions. It is, however, not without its criticisms. Some consider the DSM as a reductive approach to symptomatology and diagnosis leading to the pathologisation of everyday life. There is an alternative to the DSM grounded in psychodynamic theory. It is called the Psychodynamic Diagnostic Manual and is now in its second edition (PDM-2). The latter is an attempt to move away from conceptualising mental health disorders as a series of symptom lists which is often deficit-based, to a clinical conceptualisation of the entire individual. The PDM-2 serves to provide an alternative to the DSM and the World Health Organization's International Classification of Diseases (ICD) corpora.

dlPFC: see **dorsolateral prefrontal cortex**

dorsolateral prefrontal cortex: located in the front part of the front brain, the dorso (meaning 'back,' as in dorsal fin) lateral (meaning 'side') PFC is the brain's 'timekeeper.' It is responsible for helping a person to identify that they are in the present. This region has been described as responsible for 'cold cognition'—unemotional, factual, and attentive.

DSM-5: the fifth edition of the DSM, published in 2013 and, as mentioned previously, revised in 2022 (DSM-5-TR). Although it is difficult for medical professionals and researchers to agree on the number of disorders within the DSM—some disorders present as distinct conditions (e.g. Bipolar Disorder) while others represent various manifestations of the same illness (e.g. Cocaine Use Disorder, Cocaine Intoxication, and Cocaine Withdrawal)—one estimate is that the DSM-5 includes ~157 mental maladies (or disorders) compared to ~60 when it was first published in 1952. In any case, the diagnostic list increases with each edition, as seen with the DSM-5-TR, which included, among others, Prolonged Grief Disorder to describe individuals who mourn over a loved one for longer than one month, causing significant psychosocial distress in the process.

dual awareness: a process whereby a person focuses on two or more things simultaneously. Like now, if you're reading this while paying attention to something in the background, that could be considered dual awareness. Some might call this multi-tasking. However, neuroscientists who have studied multi-tasking have found the brain cannot multitask but only shift attention from one thing to another relatively quickly. Refer to the article below for more context: https://neurosciencenews.com/multitasking-brain-mechanism-6949/

EFT: see **Emotional Freedom Techniques**

electrochemical activity: the process whereby a neuron sends electrical impulses down the axon to activate the lower terminals, releasing various chemicals called neurotransmitters into the synapse (the small gap between one neuron and the next). This process activates the following neuron and continues electrochemical communication. It is referred to as electrochemical because both electrical impulses *and* chemical secretions deliver messages around the body.

EMDR: see **Eye Movement Desensitization and Reprocessing**

Emotional Freedom Techniques: also known as tapping or EFT. EFT was the name given to a tapping technique Gary Craig learned from his teacher Roger Callahan (1925-2013). EFT includes a process where one taps on 'meridian points' purported to be scattered throughout the body while identifying a stressor and reciting self-affirmation statements. This technique has been demonstrated to reduce limbic stress and cortisol release as measured through oral swabs before and following the procedure. EFT can be considered a dual awareness treatment approach that benefits the person in a similar way that RET and EMDR do.

engram: a neuropsychological concept contending that cognitive information becomes imprinted upon a physical substrate, for example, a memory imprinted in the hippocampus that leads to biophysiological changes. Neuroscientists understand that experiences change the brain. Engrams are cognitive bits of experience with encoded correlates within various parts of the brain.

extrapyramidal system: a part of the lower brainstem that controls involuntary motor movements such as reflexes; general motor movements such as walking, running, swimming; and posture.

Eye Movement Desensitization and Reprocessing: an eight-phase psychological treatment approach where clinicians: 1) take a client's history; 2) educate the client about what follows in treatment; 3) undertake a formal assessment of presenting issues; 4) engage in a dual-awareness desensitisation process (e.g. eye movements or other forms of bilateral stimulation such as the alternating sounding of tones in earphones or tapping alternate sides of the body); 5) 'install' new thoughts/feelings about the initial memory; 6) conduct a body scan; 7) close treatment; and 8) re-evaluate for outcomes.

fall response: a part of the human stress response whereby the body attempts to preserve energy by slowing down basic functions (e.g. breathing rate, heart rate, digestion). This response occurs when a person cannot escape a physical (or, for some humans, emotional) threat. I call this the body's last-ditch effort to protect itself when the fight-flight-freeze responses are insufficient. This part of the stress response can appear dissociative (e.g. making one feel like they are outside of or disconnected from their own body). As with the freeze response, endogenous opioids are released in excess, and individuals might feel no pain while in this state. There have been accounts of people having been attacked by animals and, although conscious (though limp), cannot feel the pain of the attack—see Peter A. Levine's book *In An Unspoken Voice* for more. Clinically, this can present as a sinking feeling that makes one fold into oneself or adopt a fetal position. It can also manifest in slouching shoulders, a forward-leaning posture, a downcast disposition, loose body posture, lethargy, a glazed-over look, and daydreaming. Imagine a sinking feeling that makes you fold into yourself and feel like your energy is be-

being drained.

fight-or-flight: a biological stress response that prepares the body to tackle a threat. An organism can either attack (fight) or get out of there (flight). Put simply: the fight response happens from the waist up; the flight response happens from the waist down. A person who is a natural fighter might feel their jaw tighten, shoulders rise, fists clench, and breathing become shallow. A person more likely to avoid danger might feel their legs become restless, look about for an exit, tremble from the waist down, or feel their knees begin to knock.

fMRI: functional magnetic resonance imaging is a brain and body imaging technique that measures blood flow as it travels within various body parts. Where there is blood flow, there is cellular activity. In the brain, blood flow occurs alongside electrochemical activity. fMRI can measure the brain's neural activity when a person is performing a particular task (compare with CT scan).

freeze response: what might be a camouflage response in the wild becomes a 'let's wait it out and hope it goes away' response in human beings. In stressful situations, the body can freeze up, becoming tense, shaky, and unable to move—think fight/flight happening simultaneously. In an argument, this might look like a person is frozen, unsure what to say, stammering, or becoming rigid. In high-anxiety individuals, this might look as though they are staring at you without blinking, their body barely moving, rigid. The neuroscientist Andrew Huberman highlights a structure within the midbrain—the periaqueductal gray—as central for the freeze response, releasing a cascade of endogenous opioids that can lead to calm and reduced pain perception.

glial cells: considered 'helper cells' within the nervous system, they support neurons in a variety of ways. ***Oligodendrocytes*** produce myelin sheath to insulate a neuron, facilitating smoother and quicker communication between cells. ***Astrocytes*** monitor chemicals that enter the brain through the bloodstream; ***ependymal cells*** produce spinal fluid, cushioning the brain and protecting it from injury; ***radial glia*** are stem cells which means they are responsible for producing new neurons and even forming other glial cells such as astrocytes and oligodendrocytes; ***microglia*** detect and remove damaged cells. Glial cells make up half the number of cells in the human brain (about 80 billion); neurons make up the rest (also, roughly 80 billion).

Hebbian principle: also referred to as 'Hebbian learning', this principle was proposed by the neuropsychologist Donald Hebb (1904-1985) and argues that *neurons that fire together, wire together.* In Hebb's own words: "When an axon of cell A is near enough to excite cell B and repeatedly or persistently takes part in firing it, some growth process or metabolic change takes place in one or both cells such that A's efficiency, as one of the cells firing B, is increased." Combining behavioural psychology with his study of the human brain, Hebb argued that with each neural impulse, a subsequent neuron becomes activated and, when repeated enough, the connection strengthens, making it more likely the same activity would occur in the future. Hebb first wrote about this in his 1949 book *The Organization of Behavior: A Neuropsychological Theory*.

hippocampi: see **hippocampus**

hippocampus: a brain structure located deep within the temporal lobes and limbic system responsible for learning and memory. The hippocampus receives input from the cerebrum (outer layer of the brain) and is, in part, involved in the process of creating memories about things that have happened in the past (episodic memories).

hypothalamus: located beneath the thalamus just above the roof of your mouth and responsible for automatic somatic processing, including preparing and responding to danger. This 'watch guard' tells other parts of the brain what to do and which chemicals/hormones need to be signalled for secretion when under threat or needing to react quickly.

insula: located deep within one of the cerebral folds (the lateral sulcus [Latin for 'groove']), separating the parietal and temporal lobes. It is a small cortical region responsible for social and emotional processing, visceral functions, and even cognitive processing. When in a conversation with another person, notice how your body reacts to the conversation as you decide what to do or say—this is the insula in action.

left hemisphere: of the brain is associated, among other things, with language, rational processing, analytical comprehension, and making sense of things.

limbic system: a midbrain region located beneath the cerebrum and above the brain stem, responsible for emotional activation, fear, memory, and attention. It is comprised of numerous brain areas such as the ***olfactory bulb*** (plays a vital role in smell), ***amygdala*** (fear, anxiety, response activation), ***hippocampus*** (learning and memory) and the ***cingulate gyrus*** (assists in the formation of memories and links them with emotions).

medial prefrontal cortex: located in the middle of the PFC and responsible for integrating the instinctive brain stem, limbic system, and cortex. Also responsible for balancing the sympathetic and parasympathetic branches of the autonomic nervous system. This allows for emotional attunement and empathy, a balanced response to stress, intuition, and fear extinction.

meridian: in Chinese medicine, the meridian system represents a conglomerate of areas throughout the human body (e.g. sides of the eyes, forehead, under the nose, chin, chest, and so on) that carry 'life energy' and, if activated, purportedly assist with attenuating illness. So far, no scientific consensus exists about their actual reality in the real world.

mesolimbic pathway: referred to as the 'reward circuitry' that activates dopamine release. The mesolimbic pathway travels from the midbrain (just above the brainstem) to the forebrain (within the frontal lobes). It reinforces learning (positive and negative) through the activation of the dopamine system. Many associate dopamine with pleasure, excitement, and reward. However, the primary function of dopamine in the nervous system is movement—in particular, goal-oriented and survival-based movement. For example, even various sea worms secrete dopamine when they detect food in their environment. People with Parkinson's disease experience a malfunction in the production of dopamine, thus leading to tremors and other movement disorders. When one feels motivated towards a particular action, dopamine is released. With dopamine, it is as much about anticipation as it is about the perceived outcome.

mPFC: see **medial prefrontal cortex**

neural firing: electrochemical activation between neurons within the synapse.

neural networks: see **neural pathways**

neural pathways: described highways of interconnected neurons grouped together depending on the function and the strength of their connections. The more these neurons are activated, the stronger that link becomes. Refer to the **Hebbian principle** for more.

neurobiological: originating from the field of neurobiology, this discipline combines the study of the nervous system with the systematic study of the human body. It seeks to understand causes and effects, conceptualising human behaviour in biologically and evolutionarily sound terms.

neurogenesis: refers to the formation of new neurons in the brain. Radial glia are the stem cells responsible for neurogenesis.

neurons: brain cells. The common identifying features of every brain cell is the ***cell body*** (soma—the nucleus and DNA are kept here), ***dendrites*** (branch-like extensions that protrude from the soma to receive information from other neurons), ***axons*** (extensions protruding from the cell body and leading to the terminals at the end), ***myelin sheath*** (insulating protection wrapped around the outer layer of the axon, helping with the speed and ease of transmission of electrochemical messages), and the ***axon terminals***

(bulb-like caps on the branch-like axon endings). If you picture a tree (both what you see above and below the ground), neurons look approximately like this.

neuroplasticity: a term that describes the way neurons evolve as a result of human experience. This process is achieved through: 1) ***neurogenesis*** (the formation of neurons); 2) ***synaptogenesis*** (the formation of synapses); 3) ***axonal sprouting*** (the formation of axons); 4) ***dendritic evolution*** (the formation and change within dendritic connections); and 5) ***synaptic pruning*** (the loss of connections between brain cells). These processes tend to be most active in areas like the olfactory bulb (a brain region important for smell) and the hippocampi. In particular, neuroplasticity occurs in relation to the long-term potentiation of cells (i.e. where synapses are strengthened for the transmission of electrochemical messages).

nervous system: the most important body system that makes up the human organism. This system is comprised of the central and peripheral nervous systems, further broken down into their subsequent parts. The ***central nervous system*** comprises the brain and spinal cord. The ***peripheral nervous system*** houses nerves that protrude from the cranium and spinal cord, branching out to the rest of the body. The peripheral nervous system is further broken down into the somatic and autonomic nervous systems. The ***somatic nervous system*** allows for voluntary movements; the ***autonomic nervous system*** controls involuntary movements and is further broken down into the ***sympathetic nervous system*** (gets the body ready for action), ***parasympathetic nervous system*** (calms the body down following action), and ***enteric nervous system*** (located around the digestive tract, controlling for gastrointestinal functions/reflexes and the release of neurotransmitters like acetylcholine, serotonin, and dopamine). The latter is also referred to as 'the second brain'.

nucleus accumbens: a region of the human brain located within the mesolimbic pathway which connects the midbrain to the limbic system and forebrain. These clusters of brain cells help mediate desire and reward through the release of dopamine and the reinforcement of positive experiences. When you've worked hard on something despite its difficulty, you will observe this region becomes activated in a brain imaging scan.

occipital lobe: an area at the back of the brain responsible for discerning visual input. Interestingly, people with damage in this area cannot see despite having perfectly functioning eyes. This is because even though their eyes work perfectly, their brain cannot process the information coming in. People with this condition might still respond to the external environment but be unaware of why they are doing so. For example, they may walk into a room and avoid crashing into the chairs and tables without being able to see them. Or, they might respond with fear or concern when shown images of angry faces but not be able to describe why. This phenomenon is called 'blindsight'.

oFC: see **orbitofrontal cortex**

orbitofrontal cortex: located within the PFC at the very front of the brain just above the eye sockets. This area is responsible for visual recognition, impulse control, and decision-making specific to reward and punishment. For example, if one sees a $100 note blowing with the wind, he may decide to pursue it. However, they must decide whether to take the risk or let it pass if it blows onto the road and a car is coming. This is the oFC in action. An important study on the relationship between the orbitomedial PFC (geographically close to the oFC) found evidence that the "hippocampal–orbitomedial prefrontal cortex circuit may be involved in attentional monitoring of the internal sensorium. This attentional monitoring system [...] is the working memory of viscero-emotional processing. [This circuit] can thus be viewed as a discrepancy detector [...] [signalling ...] attentional systems to activate, process, and reconcile internal (past) with external (present) environments, and [...] alter active working emotional 'sets' to exert cognitive–emotional control of behavior." Essentially, these brain regions allow people to pay attention to what their body is saying and think of the past within the present, choosing to act in a way that's best. More on this can be found in Wall and Messier's (2021) paper: *The hippocampal formation — orbitomedial prefrontal cortex circuit in the attentional control of active memory.*

organ system: see **body system**

peripheral nervous system: a part of the human nervous system that includes the nerves that protrude from the brain (called cranial nerves) and the spinal cord, along with their branches extending throughout the body. These nerves communicate messages from the brain to the rest of the body and take messages from the body and communicate them to the brain. Messages are conducted through nerve impulses carried by nerve cells (neurons) and transmitted through the synapse.

PFC: see **prefrontal cortex**

pituitary gland: a gland about the size of a pea hanging underneath the thalamus, like an apple, that secretes hormones like oxytocin (responsible for childbirth and human connection), growth hormone, prolactin (helps with breastfeeding), and adrenocorticotropic hormone (stimulates the release of the stress hormone cortisol). It is often called 'the Master Gland'.

Polyvagal Theory: a theory posited by the scientist and professor of psychiatry Stephen Porges which proposes that the 10th cranial nerve (the ***vagus nerve***) helps keep many major body systems (particularly the ***cardiovascular***, ***musculoskeletal***, ***gastrointestinal***, and ***sympathetic and parasympathetic nervous systems***) in check. This wandering nerve—vagus in Latin means 'wanderer'—regulates the stress response and pushes the body to respond appropriately (consider the ***fight-flight-freeze-fall*** responses). Importantly, Porges identified an important factor within the stress response to be the drawing close to trusted others (this is activated, in theory, by the ventral vagal system, leading to what Porges calls neuroception—the ability to meaningfully connect with others at a neural level). This response is evident in children who draw close to their caregivers when they feel shy, stressed, or unsure. In Porges's own words, "The polyvagal theory describes an autonomic nervous system that is influenced by the central nervous system, sensitive to afferent [sensory] influences, characterized by an adaptive reactivity dependent on the phylogeny [evolutionary history of a species] of the neural circuits, and interactive with source nuclei in the brainstem regulating the striated [parallel or groove-like] muscles of the face and head. [...] Its specific focus is on the phylogenetic shift between reptiles and mammals that resulted in specific changes to the vagal pathways regulating the heart. As the source nuclei of the primary vagal efferent [motor] pathways regulating the heart shifted from the dorsal motor nucleus of the vagus in reptiles to the nucleus ambiguus in mammals, a face–heart connection evolved with emergent properties of a social engagement system that would enable social interactions to regulate visceral state." See Porges's (2011) paper: *The polyvagal theory: New insights into the adaptive reactions of the autonomic nervous system.*

posttraumatic stress disorder: a disorder first identified in 1980 within the DSM-III following Vietnam War veterans' return back to the U.S. Many of these veterans suffered severe mental distress associated with things they had done or seen done during the war. Impacts of the affliction were exacerbated by their adversarial experiences upon returning home—this was the first war documented and broadcast in detail, leading to major protests. Before it was classified as a mental disorder, it was called ***shell shock***, ***combat fatigue***, and ***war neurosis***. PTSD is limited in diagnostic applicability given it precludes some severe stresses (e.g. bullying, viewing a traumatic video, suffering distress about something you have done to somebody else, abandonment and neglect in children, and learning about something traumatic that has occurred that involved you or somebody close to you) as they don't fall within Criterion A diagnostic events (i.e. death,

death threats, actual or threatened serious injury, or actual or threatened sexual violence). PTSD appears in the ***DSM-5-TR***, ***ICD-11***, and ***PDM-2***. However, ***complex PTSD*** (C-PTSD) only appears in the ICD-11 and PDM-2. C-PTSD (also called ***developmental trauma***) refers to severe and enduring personality, behavioural, experiential, and neurological changes following severe and prolonged traumatic stress resulting from childhood abuse and neglect. Importantly, in C-PTSD, abuse need not be physical or sexual but can include psychological abuse, neglect, feeling chronically unsafe, and abandonment.

prefrontal cortex: a brain region located just between your eyes and within the frontal lobes, including various structures responsible for higher-level functions such as self-awareness, moral judgement, meaning-making, insight, response flexibility, empathy, and self-regulation.

psychodynamic tradition: beginning in the late 1800s with the neurophysiologist Josef Breuer (1842-1925) and his young student Sigmund Freud (1856-1939). This tradition, although started by Breuer with his treatment of Anna O (i.e. Bertha Pappenheim [1859-1936]), utilised the novel approach of speaking anything that came to mind, a process he called ***the talking cure***. At the time, Pappenheim referred to this process as 'chimney sweeping' and reported that it effectively reduced emotional reactivity associated with stressful events. The talking cure was carried to its fulfilment by Freud and his clinical theories. Freud and Breuer published a book in 1893 called *Studies in Hysteria* which described the roots of psychological trauma and its effects on a person's psychology. The psychodynamic tradition posits that events during one's childhood greatly influence adulthood and shape one's personality. The main tenet of the psychodynamic tradition is that people are controlled by unconscious processes (emotions, reactions, responses, traumas, drives) and develop defence mechanisms to combat these. Therapy aims to explore these early-life traumas, discussing them at length and identifying the hold they have on you. This process ultimately leads to psychological healing.

PTSD: see **posttraumatic stress disorder**

Radical Exposure Tapping: developed in 2010 by the family therapist Laurie MacKinnon, RET integrates various aspects of EMDR and EFT, allowing for a more versatile technique to achieve similar results in a time-limited manner. It is ***radical*** because it can effect change in as little as one session. It is an ***exposure***-based therapy as it uses the ***tapping*** protocol from EFT and integrates it within the EMDR protocol (i.e. identifying sensory, somatic, cognitive, and distress-based elements of a particular event). The clinician incorporates the EFT tapping procedure into the protocol to replace bilateral stimulation until subjective units of distress reduce and the memory has been reprocessed. The full RET process involves: 1) completion of a genogram; 2) completion of a chronological trauma memory list; 3) completion of a chronological positive memory list; 4) engagement in a mindfulness exercise (i.e. safe place exercise); 5) administration of the 21-item depression, anxiety, and stress scales (DASS-21); 5) administration of the 20-item PTSD checklist for the DSM-5 (PCL-5) to measure symptoms of trauma, anxiety, stress, and depression; 6) administration of the RET protocol for each memory the client wishes to address, preferably beginning with the earliest negative event; 7) re-administration of the RET protocol once SUD reduce to 0-1 in order to 'install' positive self-statements; and, 8) rescoring of the DASS-21 and PCL-5 to measure relevant changes.

rapid eye movements: refers to the action that takes place both when one is asleep—REM sleep is also called 'dream sleep' because when a person is awoken as they enter this fifth phase of sleep, they report dreaming about 80% of the time—and during the EMDR protocol. Some theorists believe it is REM that accounts for the change process.

relaxed breathing: a relaxed breathing rate is usually about 10-12 breaths/minute. Anything higher is usually considered shallow breathing, meaning the body's sympathetic nervous system has become activated. Many breathing techniques slow a person's breathing rate and induce a more relaxed parasympathetic response.

REM: see **rapid eye movements**

RET: see **Radical Exposure Tapping**

Rogerian psychotherapy: commonly referred to as 'person-centred therapy,' this approach was first developed by the renowned psychotherapist and researcher Carl Rogers (1902-1987) in the 1950s. Rogers identified various core conditions required for successful therapy: ***congruence*** (where the therapist is honest and true to themselves, not putting up a façade); ***unconditional positive regard*** (where the therapist accepts where the client is at that moment without judging them or expecting them to be achieving more than what they are able to); and ***empathy*** (where the therapist aims to see things from the other's perspective while withholding judgement).

safe place exercise: an imagery-based mindfulness exercise that guides a person into finding an image of a place (either real or imagined) in his or her mind where they feel safe and relaxed. Sensory experiences—visual, olfactory, auditory, and somatic—are explored when the person 'visits' this place. For example, the therapist might say something like: *"Picture a place, any place, you feel calm, safe, and relaxed. It might be a place you've visited before, a place from your childhood, or a place in your own mind, a pl-*

ace you've seen in books or movies—any place that's safe, relaxing, peaceful, where you feel at ease, in control. Can you describe this place to me? What you see close by, far off in the distance, anything at all. Are there any sounds you notice? Smells? Do you notice any taste distinctive of this place? How do you feel, in your body, as you picture this place?" Most people can access a place like this quite quickly. Others take a bit of time and probing. Some cannot access a place at all (these are usually people who might have been severely traumatised or are dissociative).

schemas: core beliefs about oneself, others, and the world at large. These beliefs are often implicit, below conscious awareness. They are usually emotional, felt deeply, and are considered as true not just for you but the world at large. Common trauma beliefs include: "The world is a dangerous place", "Everyone's out to get you", "Everybody just thinks about themselves; they don't care about you", "You'll never be safe", "You've always gotta watch your back." These implicit beliefs can change the physical structure of the brain and associated neural networks and, when accessed, they become the lens through which one sees the world.

SPECT imaging: a nuclear imaging technique (i.e. a scan that picks up radiation within the body after a radioactive substance is consumed by the patient). The images taken by SPECT are projected onto a screen, showing blood flow in the corresponding part of the body. Just like fMRI, blood flow denotes activity. SPECT can pick up on areas in the body that are active compared to inactive. Compared to fMRI, SPECT produces an image representing 10-15 minutes of activity, while fMRI produces a second-by-second image.

subjective units of distress: a scale that was developed by psychiatrist Joseph Wolpe (1915-1997) measuring distress from 0 (no distress at all) to 10 (the most distressed I've ever felt).

SUD: see **subjective units of distress**

sympathetic arousal: activation of the sympathetic nervous system. The heart picks up, breathing accelerates, hands get sweaty, eyes become focused, legs are energised, and so on. This can be either positive (as with sport, excitement, or sex) or negative (as with trauma, anxiety, or distress).

sympathetic nervous system: a part of the autonomic nervous system that is involved when the body moves towards a particular action orientation. This usually involves activating (or deactivating) the major body systems (cardiovascular, respiratory, nervous, endocrine, digestive) to utilise energy and/or protect the organism from danger.

synaptic pruning: a process whereby the brain eliminates unneeded neural connections.

As children mature into their teens and twenties, their interests become more specific, allowing for the shedding of under-utilised neural circuits. Synaptic pruning is amplified during adolescence (roughly between 12 and 24 years). This part of the neuroplastic process is instrumental in the specialisation of a particular action or habit.

synaptogenesis: refers to the process by which new neural connections are formed through synapse-formation. This process is linked to synaptic pruning and the Hebbian principle.

thalamus: a region of the human brain located atop the hypothalamus and brainstem. This region is responsible for sifting and sending sensory data (sight, sound, feel, smell) to where it is needed most. Because of its location (between the midbrain, atop the brainstem, and underneath the cerebral cortex), it is well-positioned to distribute relevant information from all these regions, allowing a person to make sense of it all. For example, if you are easily frightened, you may notice the hair on the back of your neck pricks up, your heart beats faster, and you are more attuned to the noises in your surroundings (e.g. the wind blowing through the window) as you watch a horror movie. All this sensory information is relayed through the thalamus.

theory of imaginal exposure: this theory proposes change through the gradual exposure of psychologically noxious stimuli until a person is desensitised. The more one is exposed to something frightening the less anxious they are when gradually exposing themselves to it. Consider a child scared to go down a ramp at their local skate park but builds the confidence to do so over time.

Thought Field Therapy: the precursor to EFT, TFT is an 'energy-based therapy' that utilises complex tapping algorithms of various meridian points purportedly located throughout the body to facilitate psychological change. Unlike EFT, TFT posits a plethora of tapping algorithms that work for different maladies (fear, anxiety, pain, phobias, etc.).

trauma: *any* perceptively inescapable event that is perceived as overwhelming at the time of its happening and causes lasting psychological, neurobiological, physiological, interpersonal, and spiritual (or moral) problems long after it has ceased.

trauma re-processing therapy: *any* therapy aimed at redressing a traumatic incident by revisiting it in a systematic, stepwise way. These therapies explore traumas by breaking them down, utilising a technique to reduce affect, and revisiting the original memory afterwards to measure whether the response is still triggering.

ventrolateral prefrontal cortex: an area of the PFC at the intersection of the ventral (bottom) and lateral (side) areas. It is responsible for taking back cognitive control to make important decisions in a given time and place, especially those involving emotions, memories, and social and ethical behaviours (i.e. doing the right thing in a given circumstance). Think of the last time you were around others, drinking alcohol, and said something you know you shouldn't have. This would be an example of when this part of the brain was offline.

ventromedial prefrontal cortex: an area of the PFC at the intersection of the mid-bottom junction. This part of the brain accounts for emotional decision-making, communicating closely with the nucleus accumbens which helps mediate desire and reward through the release of dopamine and the reinforcement of positive experiences. There is some cross-over between the functions of the vmPFC and the vlPFC.

vlPFC: see **ventrolateral prefrontal cortex**

vmPFC: see **ventromedial prefrontal cortex**

working memory: refers to the cognitive space that allows for the holding and switching between short-term verbal, visual, and attentional information. Working memory helps keep multiple things in mind simultaneously, such as remembering a number long enough to input it into your phone, and is thought to be mediated by the PFC.

working memory hypothesis: the theory positing that one can only balance a certain number of items simultaneously in working memory. It is thought that there is only a short window of time for a memory to be altered once it is recalled—the working memory hypothesis proposes that if one 'overloads' working memory through a dual awareness process, the reactive and emotive elements of that memory can reduce in intensity and allow for the introduction and consolidation of new information.

IMAGES & ENDNOTES

Fig. 1

Author: Jennifer Walinga. Source: https://opentextbc.ca/introductiontopsychology/chapter/3-1-the-neuron-is-the-building-block-of-the-nervous-system/. Used with permission under the Creative Commons Attribution-Share Alike 4.0 International license.

Fig. 2

Author: Anton J. M. Loonen and Svetlana A. Ivanova. Source: https://www.frontiersin.org/articles/10.33 89/fnhum.2016.00571/full. Used with permission under the Creative Commons Attribution 4.0 International license.

Fig. 3

Author: CNX OpenStax. Source: https://openstax.org/books/biology/pages/1-introduction. Used with permission under the Creative Commons Attribution 4.0 International license.

Fig. 4

Author: OpenStax & Tomáš Kebert & umimeto.org. Source: https://commons.wikimedia.org/wiki/File: Endocrine_English.svg. Used with permission under the Creative Commons Attribution-Share Alike 4.0 International license.

Fig. 5

Author: Mobbs D, Lau HC, Jones OD, Frith CD. Source: https://journals.plos.org/plosbiology/article?id=10.1371/journal.pbio.0050103. Used with permission under the Creative Commons Attribution 2.5 Generic license.

Fig. 6

Author: Mobbs D, Lau HC, Jones OD, Frith CD. Source: https://journals.plos.org/plosbiology/article?id=10.1371/journal.pbio.0050103. Used with permission under the Creative Commons Attribution 2.5 Generic license.

Fig. 7.

Author: James.mcd.nz. Source: self-made - reproduction of combined images Surfacegyri.JPG by Reid Offringa and Ventral-dorsal streams.svg by Selket. Used with permission under the Creative Commons Attribution Alike 4.0 International license; 3.0 Unported license; and 2.0 Generic license.

Fig. 8.

Source: OpenStax College. Source: https://openstax.org/apps/archive/20220801.182716/resources/f2ed e825d615044873eaaca9adb23d939c79ff32. Used with permission under the Creative Commons Attribution 2.5 Generic license.

Fig. 9

Author: BruceBlaus. Source: Own work (expanded library of Blaus's medical imagery can be found https://en.wikiversity.org/wiki/WikiJournal_of_Medicine/Medical_gallery_of_Blausen_Medical_2014). Used with permission under the Creative Commons Attribution 3.0 Unported license.

Fig. 10

Ibid.

Fig. 11

Author: Henry Gray (1827-1861). Source: Bartleby.com: Gray's Anatomy, Plate 726. Public domain.

Fig. 12

Ibid., Plate 727. Public domain.

Fig. 13

Author: Oscar Arias-Carrión, Maria Stamelou, Eric Murillo-Rodríguez, Manuel Menéndez-González & Ernst Pöppel. Substantially modified by Wikipedia user Seppi333. Source: Ibid., Dopaminergic reward system: a short integrative review International Archives of Medicine 2010, 3:24 doi:10.1186/1755-7682-3-24. Used with permission under the Creative Commons Attribution 2.0 Generic license.

Fig. 14

Refer to attribution in ***Fig. 9***.

Fig. 15

Ibid.

Fig. 16

Ibid.

Fig. 17

Ibid.

Fig. 18

Ibid.

Fig. 19

Refer to attribution in ***Fig. 11***. Source: wikimedia.org/wiki/File:Gray727_cingulate_gyrus.png. Public domain.

[1] Duncan, B. L., Miller, S. D., Wampold, B. E., Hubble, M. A., (Eds.) (2010). *The Heart and Soul of Change: Delivering What Works in Therapy* (2nd ed.). American Psychological Association.

[2] Although I am trained in RET, I differentiate between the therapy as a whole and the RET protocol as referenced it in this book. This is because the protocol is but one part of the therapy as developed by Laurie MacKinnon. For context, refer to Barna, E., & O'Hara, D. (2019). Applying the radical exposure tapping protocol in treating trauma-related emotional reactivity: A phenomenological case study. *Psychotherapy and Counselling Journal of Australia*, *7*(1). Retrieved from http://pacja.org.au/?p=5031. The latter article can be read alongside MacKinnon's article on RET as a whole. See Note [18] for more. To date, Barna and O'Hara's (2019) article is the most recently published on the RET protocol. Importantly, within this book, I draw from elements of RET, EFT, EMDR, and other therapies. I am not the originator of any one of these. I draw from many therapeutic modalities in my clinical practice, as any effective clinician must, and will refer to those I draw from where relevant.

[3] Shapiro's experience has been reproduced in various sources. I decided upon the account appearing in the following citation as it represents the earliest representation: Shapiro, F. (1989). Efficacy of the eye movement desensitization procedure in the treatment of traumatic memories. *Journal of Traumatic Stress*, *2*(2), 199-223. https://doi.org/10.1002jts.2490020207

[4] Shapiro, F. (1997). *EMDR: The breakthrough therapy for overcoming anxiety, stress and trauma.* New York: Basic Books.

[5] This explanation of EMDR appears in my 2019 article, coauthored with Dr O'Hara. Refer to Note [2].

[6] The Australian Psychological Society (2010). *Evidence-based psychological interventions in the treatment of mental disorders: A literature review* (3rd ed.). Retrieved from http://psychology.org.au/Assets/Files/Evidence-Based-Psychological-Interventions.pdf

[7] Foa, E., Keane, T. M., Friedman, M. J., & Cohen, J. A. (2009). *Effective treatments for PTSD: Practice guidelines of the international society for traumatic stress studies* (2nd ed.) New York: Guilford.

[8] Phoenix Australia – Centre for Posttraumatic Mental Health (2013). *Australian guidelines for the treatment of acute stress disorder & posttraumatic stress disorder.* Retrieved from http://phoenixaustralia.org/wp-content/uploads/2015/03/Phoenix-ASD-PTS

D-Guidelines.pdf

[9] World Health Organization (2013). *Guidelines for the management of conditions specifically related to stress*. Geneva, Switzerland: Author.

[10] Ortner, N. (2013). *The tapping solution: A revolutionary system for stress-free living*. U.S.: Hay House.

[11] Callahan, R. (1985). *The Five Minute Phobia Cure*. Wilmington, DE: Enterprise Publishing.

[12] Callahan R. A thought field therapy (TFT) algorithm for trauma: A reproducible experiment in psychotherapy. Paper presented at the annual meeting of the American Psychological Association, New York, August 1995.

[13] Craig, G. (n.d.). *The EFT Manual* (6th ed.).

[14] Church, D. (2013). Clinical EFT as an evidence-based practice for the treatment of psychological and physiological conditions. *Psychology*, *4*(8), 645-654. https://doi.org/10.4236/psych.2013.48092

[15] Feinstein, D. (2012). Acupoint stimulation in treating psychological disorders: Evidence of efficacy. *Review of General Psychology*, *16*(4), 364-391. https://doi.org/10.1037/a0028602

[16] McCaslin, D. (2009). Comments and rejoinder: A review of efficacy claims in energy psychology. *Psychotherapy Theory, Research, Practice, Training*, *46*(2), 249-256. https://doi.org/10.1037/a0016025

[17] Refer to Note [2]

[18] MacKinnon, L. (2014). Deactivating the Buttons: Integrating Radical Exposure Tapping with a Family Therapy Framework. *Australian and New Zealand Journal of Family Therapy, 35*(3). https://doi.org/10.1002/anzf.1070

[19] Refer to Note [18]

[20] Refer to Note [2]

[21] Shapiro, F. (Ed.). (2001). *Eye movement desensitization and reprocessing (EMDR)* (2nd ed.). New York, NY: Guilford Publications.

[22] Shapiro, F. (2002). EMDR treatment: Overview and integration. In F. Shapiro (Ed.), *EMDR as an integrative psychotherapy approach: Experts of diverse orientations explore the paradigm prism* (pp. 27-55). Washington, DC: American Psychological Association.

[23] van Veen, S., van Schie, K., van de Schoot, R., van den Hout, M. A., & Engelhard, I. M. (2020). Making eye movements during imaginal exposure leads to short-lived. *Journal of Behavior Therapy and Experimental Psychiatry, 67*, 1-8. 8. https://doi.org/10.1016/j.jbtep.2019

.03.001

[24] Landin-Romero, R., Moreno-Alcazar, A., Pagani, M., & Amann, B. L. (2018). How Does Eye Movement Desensitization and Reprocessing Therapy Work? A Systematic Review on Suggested Mechanisms of Action. *Frontiers in Psychology, 9*, 1395. https://doi.org/10.3389/fpsyg.2018.01395

[25] van der Kolk, B. (2015). *The body keeps the score: Brain, mind, and body in the healing of trauma.* New York, NY: Penguin Books.

[26] Hase, M., Balmaceda, U. M., Ostacoli, L., Liebermann, P., & Hofmann, A. (2017). The AIP model of EMDR therapy and pathogenic memories. *Frontiers in Psychology, 8*(1578). https://doi.org/10.3389/fpsyg.2017.01578

[27] Solomon, R. M., & Shapiro, F. (2008). EMDR and the adaptive information processing model potential mechanisms of change. *Journal of EMDR Practice and Research, 2*, 315-325. https://doi.org/10.1891/1933-3196.2.4.315

[28] van den Hout, M. A., & Engelhard, I. M. (2012). How does EMDR work? *Journal of Experimental Psychopathology, 3*(5), 724-738. https://doi.org/10.5127/jep.028212

[29] Azevedo, F. A., Carvalho, L. R., Grinberg, L. T., Farfel, J. M., Ferretti, R. E., Leite, R. E., ... Herculano-Houzel, S. (2009). Equal numbers of neuronal and nonneuronal cells make up the human brain: An isometrically scaled-up primate brain. *Journal of Comparative Neurology, 513*(5), 532-541. https://doi.org/10.1002/cne.21974

[30] Herculano-Houzel, S. (2009). The human brain in numbers: A linearly scaled-up primate brain. *Frontiers in Human Neuroscience, 3*(31). https://doi.org/10.3389/neuro.09.031.2009

[31] Carter, R., Aldridge, S., Page, M., Parker, S., Frith, C., & Frith, U. (2014). *The brain book* (2nd ed.). London, UK: Dorling Kindersley Limited.

[32] Taylor, K., Westlake, H., Ambrose, J., Hallinan, C., & Joyce, N. (Eds.) (2020). *How the brain works: The facts, visually explained* (1st ed.). London, UK: DK.

[33] Refer to Note [26]

[34] American Psychiatric Association (2013). *Diagnostic and statistical manual of mental disorders* (5th ed.). Washington, DC: Author.

[35] Refer to Note [31]

[36] Hebb, D. (1949). *The organisation of behaviour.* New York, NY: John Wiley and Sons.

[37] Keysers, C., & Gazzola, V. (2014). Hebbian learning and predictive mirror neurons for actions, sensations and emotions. *Philosophical Transactions of the Royal Society B, 369*(1644). https://doi.org/10.1098/rstb.2013.0175

[38] Shatz, C. J. (1992). The developing brain. *Scientific American, 267*, 60-67. https://doi.org/10.1038/scientificamerican0992-60

[39] Siegel, D. J. (2011). *Mindsight: The new science of personal transformation.* New York, NY: Bantam Books.

[40] Australian Bureau of Statistics (2021). *National study of mental health and wellbeing: Summary statistics on key mental health issues including the prevalence of mental disorders and the use of services.* Retrieved from https://www.abs.gov.au/statistics/health/mental-health/national-study-mental-health-and-wellbeing/2020-21 In comparison to a 2007 survey, there is, thankfully, a small reduction in mental health distress. In 2007, more than 7 million people (45% of those sampled) reported meeting the criteria for the diagnosis of a mental health condition at some point in their lives. Of those, more than three million (20%) reported meeting that criteria over the past *12 months*. Particularly interesting, at least for our purposes, was the 12-month prevalence for PTSD alone rose from 1.5% in 1997 to 6.4% in 2007. At the time, this was the highest-grossing 12-month prevalence for *any* mental health condition in Australia. Compared to 2021 outcomes—and taking COVID-19 lockdowns into account—PTSD became the *second* highest diagnosed condition (5.7%). The first, understandably, was Social Phobia (7% prevalence). For the 2007 findings, see here: Australian Bureau of Statistics (2007). *National survey of mental health and wellbeing: Summary of results.* Retrieved from https://www.ausstats.abs.gov.au/ausstats/subscriber.nsf/0/6AE6DA447F985FC2CA2574EA00122BD6/$File/National%20Survey%20of%20Mental%20Health%20and%20Wellbeing%20Summary%20of%20Results.pdf

[41] Rosenman, S. (2002). Trauma and posttraumatic stress disorder in Australia: Finding in the population sample of the Australian National Survey of Mental Health and Wellbeing. *Australian and New Zealand Journal of Psychiatry, 36*(4), 515-520.

[42] Levine, P. A. (2010). *In an unspoken voice: How the body releases trauma and restores goodness.* Berkeley, CA: North Atlantic Books.

[43] Langford, K. (2015). *Anatomy 101: From muscles and bones to organs and systems, your guide to how the human body works.* Avon, MA: Adams Media.

[44] Roelofs, K. (2017). Freeze for action: Neurobiological mechanisms in animal and human freezing. *Philosophical Transaction of the Royal Society B, 372*(1718). https://doi.org/10.1098/rstb.2016.0206

[45] Schmidt, N. B., Richey, J. A., Zvolensky, M. J., & Maner, J. K. (2008). Exploring the human

freeze responses to a threat stressor. *Journal of Behavior Therapy and Experimental Psychiatry, 39*(3), 292-304. https://doi.org/10.1016/j.jbtep.2007.08.002

[46] Siegel, D. J. (2010). *The mindful therapist: A clinician's guide to mindsight and neural integration.* New York, NY: W.W. Norton & Company, Inc.

[47] Siegel, D. J. (2012). *Pocket guide to interpersonal neurobiology: An integrative handbook of the mind.* New York, NY: W.W. Norton & Company, Inc.

[48] Refer to Note [42]

[49] Lima, A. A., Fiszman, A., Marques-Portella, C., Mendlowicz, M. V., Coutinho, E. S., Maia, D. C., ... Figueira, I. (2010). The impact of tonic immobility reaction on the prognosis of posttraumatic stress disorder. *Journal of Psychiatric Research, 44*(4), 224-228. https://doi.org/10.1016/j.psychires.2009.08.005

[50] Barna, E. (2017). *Minding the brain towards change: What willpower really means for addiction.* Melbourne, VIC: Barna Books.

[51] Refer to Note [25]

[52] Ibid.

[53] Refer to Note [2]

[54] Boals, A. (2016). Trauma in the eye of the beholder: Objective and subjective definitions of trauma. *Journal of Psychotherapy Integration, 1-13.* https://doi.org/10.1037/int0000050

[55] Briere, J. N., & Scott, C. (2015). *Principles of trauma therapy: A guide to symptoms, evaluation, and treatment* (2nd ed.). Los Angeles, CA: SAGE Publications.

[56] Levine, P. A. (1997). *Waking the tiger: Healing trauma.* Berkeley, CA: North Atlantic Press.

[57] May, C. L., & Wisco, B. E. (2016). Defining trauma: How level of exposure and proximity affect risk for posttraumatic stress disorder. *Psychological Trauma: Theory, Research, Practice, and Policy, 8*(2), 233-240. https://doi.org/10.1037/tra0000077

[58] Refer to p. 11 of Levine, P. A. (2008). *Healing trauma: A pioneering program for restoring the wisdom of your body.* Boulder, CO: Sounds True.

[59] Carter, R., Aldridge, S., Page, M., Parker, S., Frith, C., & Frith, U. (2014). *The brain book* (2nd ed.). London, UK: Dorling Kindersley Limited.

[60] Refer to Note [47]

[61] Rasia-Filho, A. A., Londero, R. G., & Achaval, M. (2000). Functional activities of the amygdala: An overview. *Journal of Psychiatry and Neuroscience, 25*(1), 14-23.

[62] Refer to Note [59]

[63] Refer to Note [39]

[64] Guggisberg, A. G., & Mottaz, A. (2013). Timing and awareness of movement decisions: Does consciousness really come too late? *Frontiers in Human Neuroscience*, *7*, 385, https://doi.org/10.3389/fnhum.2013.00385

[65] Libet, B. (1993). The neural time factor in conscious and unconscious events. *Ciba Foundation Symposium*, *174*, 123-137. https://doi.org/10.1002/9780470514412.ch7

[66] Anand, K. S., & Dhikav, V. (2012). Hippocampus in health and disease: An overview. *Annals of Indian Academy of Neurology*, *15*(4), 239-246. https://doi.org/10.4103/0972-2327.104323

[67] Kim, E. J., Pellman, B., & Kim, J. J. (2015). Stress effects on the hippocampus: A critical review. *Learning and Memory*, *22*(9), 411-416. https://doi.org/10.1101/lm.037291.114

[68] You can find the interview transcript here: https://www.jordanbpeterson.com/blog-posts/sunday-times-unedited-interview-transcript/

[69] Refer to Note [25]

[70] Refer to Notes [39] and [47]. In these references, the neuropsychotherapist, Dan Siegel, discusses the importance of prefrontal activation in therapeutic change. Essentially, for Siegel at least, it's the PFC that is front and centre for any change achieved in psychotherapy.

[71] Refer to Note [25]

[72] Siegel, D. J. (2006). An interpersonal neurobiology approach to psychotherapy: Awareness, mirror neurons, and neural plasticity in the development of well-being. *Psychiatric Annals*, *36*(4).

[73] van der Kolk, B. (2000). Posttraumatic stress disorder and the nature of trauma. *Dialogues in Clinical Neuroscience*, *2*(1), 7-22.

[74] Refer to Note [67]

[75] Lanius, R. A., Williamson, P. C., Densmore, M., Boksman, K., Neufeld, R. W., Gati, J. S., Menon, R. S. (2004). The nature of traumatic memories: a 4-T FMRI functional connectivity analysis. *The American Journal of Psychiatry*, *161*(1), 36-44. https://doi.org/10.1176/appi.ajp.161.1.36

[76] Refer to Note [25]

[77] Doidge, N (2007). *The brain that changes itself: Stories of personal triumph from the frontiers of brain science*. New York, NY: Penguin Books.

[78] Kays, J. L., Hurley, R. A., & Taber, K. H. (2012). *The dynamic brain: Neuroplasticity and*

mental health. The Journal of Neuropsychiatry and Clinical Neurosciences, 24(2), 118-124. https://doi.org/10.1176/appi.neuropsych.12050109

[79] Refer to Note [36]

[80] Refer to Note [38]

[81] Kandel, E. R. (2000, December). The molecular biology of memory storage: A dialogue between
genes and synapses. *Nobel Lecture*. Lecture presented at the 2000 Nobel Lectures, New York, USA.

[82] Refer to Note [2]

[83] Baldwin, S. A., Wampold, B. E., & Imel, Z. E. (2007). Untangling the alliance-outcome correlation: Exploring the relative importance of therapist and patient variability in the alliance. *Journal of Consulting and Clinical Psychology*, *75*, 842-852. https://doi.org/10.1037/0022-006X.75.6.842

[84] Gentry, E. J., Baranowsky, A. B., & Hoton, R. (2017). Trauma competency: An active ingredients approach to treating posttraumatic stress disorder. *Journal of Counselling & Development*, *95*(3), 279-287. https://doi.org/10.1002/jcad.12142

[85] Hess, N. (2019). A neuroscientific perspective on the therapeutic alliance and how talking changes the brain: Supporting a common factors model of psychotherapy. *Psychotherapy and Counselling Journal of Australia*, *7*(2). Retrieved from http://pacja.org.au/?p=5613

[86] Lakeman, R. (2019). Advanced empathy: A key to supporting people experiencing psychosis or other extreme states. *Psychotherapy and Counselling Journal of Australia*. Retrieved from Retrieved from http://pacja.org.au/?p=5379

[87] Refer to Note [72]

[88] Refer to Note [2]

[89] Miller, G. A. (1956). The magical number seven, plus or minus two: Some limits on our capacity for processing information. *Psychological Review*, *63*(2), 81-97. https://doi.org/10.1037/h0043158

[90] Siegel, D. J. (2002). The developing mind and the resolution of trauma: Some ideas about information processing and an interpersonal neurobiology of psychotherapy. In F. Shapiro (Ed.), *EMDR as an integrative psychotherapy approach: Experts of diverse orientations explore the paradigm prism* (pp. 85-121). Washington, DC: American Psychological Association.

[91] Siegel, D. J. (2007). *The mindful brain: Reflection and attunement in the cultivation of*

well-being. New York, NY: W.W. Norton & Company, Inc.

[92] Bush, B., Luu, P., Posner, M. I. (2000). Cognitive and emotional influences in anterior cingulate cortex. Trends in Cognitive Sciences, *4*(6), 215-222. https://doi.org/10.1016/S1364-6613(00)01483-2

[93] Rolls, E. T. (2004). The functions of the orbitofrontal cortex. *Brain and Cognition*, *55*(1), 11-29. https://doi.org/10.1016/S0278-2626(03)00277-X

[94] Baxter, M. G., Gaffan, D., Kyriazis, D. A., & Mitchell, A. S. (2009). Ventrolateral prefrontal cortex is required for performance of a strategy implementation task but not reinforcer devaluation effects in rhesus monkeys. *The European Journal of Neuroscience*, *29*(10), 2049-2059. https://doi.org/10.1111/j.1460-9568.2009.06740.x

[95] Refer to Note [91]

[96] Refer to Note [77]

[97] Church, D., Yount, G., & Brooks, A. J. The effect of emotional freedom techniques on stress biochemistry: A randomized controlled trial. *The Journal of Nervous and Mental Disease*, *200*(10), https://doi.org/10.1097/NMD.0b013e31826b9fc1

[98] Gerardi, M., Rothbaum, B. O., Astin, M. C., & Kelley, M. (2010). Cortisol response following exposure treatment for PTSD in rape victims. *Journal of Aggression, Maltreatment & Trauma*, *19*(4), 349-356. https://doi.org/10.1080/1080/10926771003781297

[99] Refer to Note [91]

[100] Scorsese, M. (Director). (2019). *The Irishma*n [Motion picture]. USA: Netflix.

[101] Wolpe, J. (1969). *The practice of behaviour therapy*. New York, NY: Pergamon Press.

[102] Refer to Note [72]

[103] Finnegan, A., Kip, K., Hernandez, D., McGhee, S., Rosenzweig, L., Hynes, C., & Thomas, M. (2016). Accelerated resolution therapy: An innovative mental health intervention to treat post-traumatic stress disorder. *Journal of Royal Army Medical Corps*, *162*(2), 90-97. https://doi.org/10.1136/jramc-2015-000417

[104] Ellis, T. E., Rufino, K. A., & Nadorff, M. R. (2017). Treatment of nightmares in psychiatric inpatients with imagery rehearsal therapy: An open trial and case series. *Behavioral Sleep Medicine*, *17*(2). https://doi.org/10.1080/15402002.2017.1299738

[105] Salgado, S., & Kaplitt, M. G. (2015). The nucleus accumbens: A comprehensive review. *Stereotactic and Functional Neurosurgery*, *93*(2), 75-93. https://doi.org/10.1159/000368279

[106] Mullally, S., & Maguire, E. A. (2013). Memory, imagination, and predicting the future: A common brain mechanism? *The Neuroscientist*, *20*(3). https://doi.org/10.1177/107385841349

5091
[107] Reddan, M. C., Wager, T. D., & Schiller, D. (2018). Attenuating neural threat expression with imagination. *Neuron, 100*(4), 994. https://doi.org/10.1016/j.neuron.2018.10.047
[108] Schacter, D. L., Addis, D. R., Hassabis, D., Martin, V. C., Spreng, R. N., & Szpunar, K. K. (2012). The future of memory: Remembering, imagining, and the brain. *Neuron, 76*(4). https://doi.org/10.1016/j.neuron.2012.11.001
[109] Refer to Note [107]
[110] Refer to Note [2]
[111] American Psychiatric Association (1980). *Diagnostic and statistical manual of mental disorders* (3rd ed.). Washington, DC: Author.
[112] Levine, P. A. (2015). *Trauma and Memory: Brain and body in a search for the living past: A practical guide for understanding and working with traumatic memory.* Berkeley, CA: North Atlantic Books.
[113] Porges, S. W. (1995). Orienting in a defensive world: Mammalian modifications of our evolutionary heritage: A polyvagal theory. *Psychophysiology, 32*(4), 301-318. https://doi.org/10.1111/j.1269-8986.1995.tb01213.x
[114] Porges, S. W. (2009). The polyvagal theory: New insights into adaptive reactions of the autonomic nervous system. *Cleveland Clinic Journal of Medicine, 76*(2), S86-S90. https://doi.org/10.3949/ccjm.76.s2.17
[115] Porges, S. W., & Dana, D. (2018). *Clinical applications of the polyvagal theory: The emergence of polyvagal-informed therapies* (Norton series on interpersonal neurobiology). New York, NY: W. W. Norton & Company.
[116] Camina, E., & Güell, F. (2017). The neuroanatomical, neurophysiological and psychological basis of memory: Current models and their origins. *Frontiers in Pharmacology, 8*, 438. https://doi.org/10.3389/fphar.2017.00438
[117] Refer to Note [112]
[118] Refer to Note [42]
[119] Parvizi, J., Rangarajan, V., Shirer, W. R., Desai, N., Grecius, M. D. (2013). The will to persevere induced by electrical stimulation of the human cingulate gyrus. *Neuron, 80*(6), 1359-1367. https://doi.org/10.1016/j.neuron.2013.10.057
[120] Source in Note [112]. In his discussion of the amCC, Levine says, "Current research [...] shows that this brain region is activated when there are stimuli of strong affective salience, whether positive or negative. It has clear neural connections with areas of the insula, amygdala, hypothalamus, brain stem, and thalamus. The aMCC, along with the ins-

ula cortex, receives primary input from sense receptors inside the body. [... and] is the only part of the cortex that can actually dampen the amygdala's fear response. [...] this circuit of thalamus, insula, anterior cingulate, and the medial prefrontal Cortex receives interoceptive information [...] and affects the preparation for action via the extrapyramidal motor system [...] the very fabric of which procedural memories are made."

[121] Uddin, L. Q., Nomi, J. S., Hebert-Seropian, B., Ghaziri, J., & Boucher, O. (2017). Structure and funct-
ion of the human insula. *Journal of Clinical Neurophysiology, 34*(4), 300-306. https://doi.org/10.1097/WNP.0000000000000377

[122] Refer to Note [112]

[123] Sotres-Bayon, F., Bush, D. E., & LeDoux, J. E. (2004). Emotional preservation: An update on prefrontal-amygdala interactions in fear extinction. *Learning and Memory, 11*(5), 525-535. https://doi.org/10.1101/lm.79504

[124] Refer to Note [112]

[125] Refer to Note [90]

[126] Young, J. E., Klosko, J. S., & Weishaar, M. (2003). *Schema therapy: A practitioner's guide*. New York, NY: Guilford Publications.

[127] Refer to Note [90]

[128] Refer to Note [112]

[129] Lacey, J. W., & Stark, C. E. L. (2013). The neuroscience of memory: Implications for the courtroom. Nature Reviews. *Neuroscience, 14*(9), 649-658. https://doi.org/10.1038/nrn3563

[130] Lee, J. L. C., Nader, K., & Schiller, D. (2017). An update on memory reconsolidation updating. *Trends in cognitive neuroscience, 21*(7), 531-545. https://doi.org/10.1016/j.tics.2017.04.006

ABOUT THE AUTHOR

Emil Barna is a psychotherapist, podcaster, writer, and founder of *Barna Counselling & Consulting* and *Barna Books*.

Having studied psychology and philosophy during his undergraduate, he chose to pursue graduate study in the fields of counselling and psychotherapy where he went on to specialise in the treatment of traumatic stress, anxiety, substance use and behavioural addictions. In his aim to provide a tailored approach to psychotherapy, he integrates various approaches to help his clients build confidence through mastery, discipline, and traverse psychological barriers to become their best selves.

Emil values cost-free resourcing to psychologically-minded individuals, and to help in this area he created *The Emil Barna Podcast* where he interviews specialists in their fields from a psychological perspective about how to become one's best self. He also keeps an active social media presence, uploading practical strategies to various platforms in order to help those who aim to build mastery in their daily lives.

In his mission to provide the best reading materials he can, Emil created *Barna Books* where he continues to build a library of fictional and non-fictional work with a psychological undercurrent. This publishing house offers stories to entertain and resources to equip those readers who would like to challenge themselves and learn more about why humans are the way that we are.

Emil is married to a very patient wife and has two children. He lives in Melbourne, Australia, and is an avid reader and Brazilian Jiu Jitsu enthusiast.

Emil Barna

BA (Psychology), Grad Dip (Counselling),
Master (Counselling and Psychotherapy)
Psychotherapist | Podcaster | Writer | Founder
Barna Counselling & Consulting
Barna Books
www.barnacc.com
emiliobarna@gmail.com

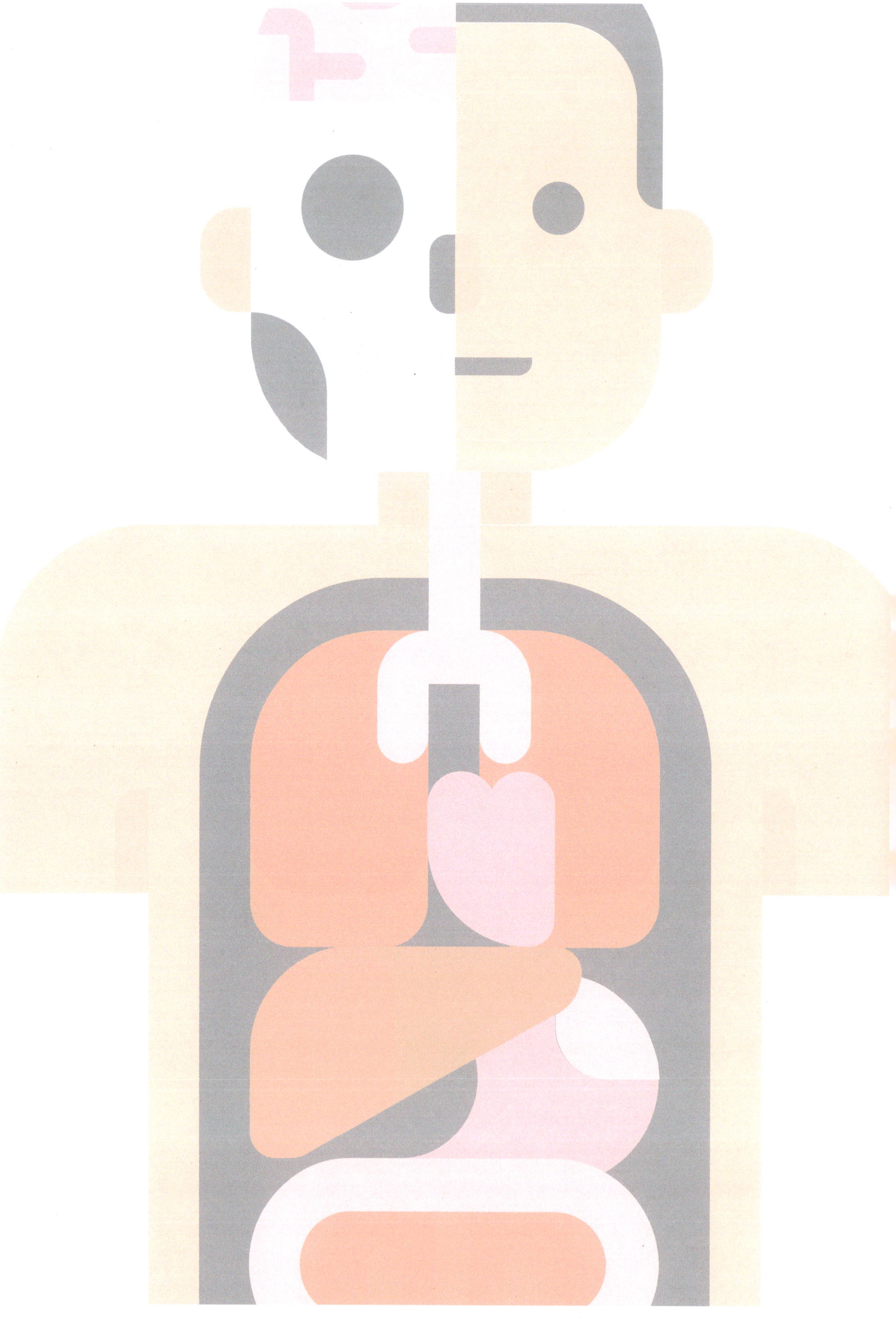

Made in the USA
Monee, IL
05 November 2026

67383934R00059